AF859083

A Geologic Guide to the Brown Canyon Area of the Baboquivari Mountains, South-Central Arizona

Early morning photo of the Baboquivari Mountains from the east. Baboquivari Peak is a granite stock. The striking pattern of rock outcrops is created by dikes of the Allison Camp Rhyolite. Courtesy of Murray Bolesta, CactusHuggers.com.

by Richard Conway, PhD

The Baboquivari range from Kitt Peak. Courtesy Fred Espenak, AstroPixels.com.

Copyright © 2016 by Richard Conway

Printed in the United States of America

ISBN 978-0-692-70317-5

Cover photo of Baboquivari Peak from the Environmental Education Center courtesy of Mary Scott, Desert Light Studio, Arivaca, Arizona.
Inside cover, photo of the Baboquivari range looking south from Kitt Peak.

Preface & Acknowledgements

Baboquivari Peak is more than a landform. It is a sacred place for the indigenous people and for many others who live in its vicinity today. It is fundamental to the culture of the Tohono O'odham Nation. Their traditions emerge from the mountains and the deserts. This is especially so with the Baboquivari Mountains. I'itoi, the Tohono O'odham Elder Brother and creator, resides in a cave within Baboquivari Peak.

Early Spanish explorers commented on " . . . the high sierra of Babuquiri . . .". European settlers came to use the land and were humbled by the mighty landscapes they chose to inhabit.

I live in Arivaca and I am always aware of Baboquivari. My decision to call this place home was a certain one as I looked across the Altar Valley to the Baboquivari Mountains and sensed a timeless and powerful presence. From my first visit to Brown Canyon in 2005 I was committed to learning about this place and sharing it with the world. This book is the fulfillment of that promise and is an act of giving back to the places and the people that have inspired me.

First among these is I'itoi. During my years exploring the canyon I'itoi was my guide and the rocks I found and the things I saw were all the more wondrous.

Mary Scott is my partner and my joy. She is a brillant source of magical power, charm and creativity with an immense love for the Universe. Together, we did great natural history retreats at the Environmental Education center in Brown Canyon. I did the rocks and Mary did the rest, to the great delight and amazement of all—including me. We had a fantastic run and had much fun doing it. Many of the remarkable photographs in this book are hers.

Next are the many visitors who participated in the programs offered in Brown Canyon. Through these adventurers, I perceived the place with new eyes and mind. I am forever grateful to them as my teachers. It was during these times that the idea of this book took form and gradually grew.

None of this would have happened without the generous and genuine support of the Buenos Aires National Wildlife Refuge staff. Brown Canyon is a gift they presented to me and this book is in appreciation.

Early in my literature research, the name Gordon Haxel appeared, the USGS geologist who is the authority on the area. When I contacted Gordon he was delighted that such a book would be written. He offered to help in any way that he could and, and between preparing a special geologic map and the challenging task of editing my writing, his contribution has been immense.

I am also happy to express my appreciation to Monica Tilley for her enthusiasm for the project and her expert editing. Phil Pearthree read the section on landscape evolution and was very helpful in teaching me about a topic that is especially compelling in this area.

Sumi-e ink drawing by Peg Bowden, 2013.

Table of Contents

Where is this Place, what is this Guide?

The Baboquivari Mountains are a range of rugged wilderness visible from much of south-eastern and south-central Arizona. Baboquivari Peak dominates the range and is a sacred place for the region's indigenous people. On the north end is Kitt Peak, a world-renowned astronomy site. South of Kitt Peak the wild and rugged crest of the Baboquivari range winds its way southward across the Mexican border. To the west is the Tohono O'odham Nation, with its Sonoran Desert ecosystem. To the east is the Altar Valley with its ranches, the Buenos Aires National Wildlife Refuge, and other public lands, most of which are wild. This is the world of I'itoi, the Tohono O'odham peoples' Elder Brother and of Macho B, the jaguar hero in a tragic tale of an encounter between the wild world and good intentions.

Tucked below Baboquivari Peak is the drainage of Brown Canyon. Most of the canyon is under the protection of the Buenos Aires National Wildlife Refuge and it is a remarkable example of a sky island ecosystem that is still intact, despite a history of ranching. Access to Brown Canyon is by arrangement with the Refuge.

The canyon south of Brown Canyon is Thomas Canyon which is a popular access to Baboquivari Peak as a hiking and climbing destination. This is where the favorite of regional literature, *The Peacocks of Baboquivari,* by Erma J. Fisk, was written.

Across the divide on the Tohono O'odham Nation is Baboquivari Park. The Baboquivari Park side and the Brown Canyon side are two different worlds, both presided over by Baboquivari Peak. It is not possible to truly discover the area without visiting both sides. Access to the

Painting of Arivaca and Baboquivari Peak by Carl Schuchard in 1857. Schuchard was a traveler and mine promoter. The image is from the Report of the Sonora Exploring and Mining Co., made to the stockholders, September, 1857, Cincinnati: Railroad Record print, 1857, p. 7.

Baboquivari Park area is easily acquired by contacting the Nation or obtaining a permit at the visitor center, located at the turnoff on Indian Route 19 to Baboquivari Park.

The first European explorers were Spanish and in 1695 the Franciscan priest, Father Eusebio Kino, established a visita at the present site of Arivaca. In 1864 Pedro Aguirre established the Buenos Aires Ranch which went through several ownerships until it was sold to the United States Fish and Wildlife Service in 1985.

Ranching in Brown Canyon began with Rolin Brown, a Tucson newspaper publisher, in the 1870's. Despite the challenges of ranching a desert mountain canyon it continued to be intermittently ranched until a coalition of owners sold it to the United States Fish and Wildlife Service in 1995.

Systematic geologic studies by the United States Geological Survey (USGS) of the Baboquivari Mountains began in the late 1950s. In the late 1970s and 1980s a major effort was initiated to document the region's geology by mapping the core map quadrangles of the Tohono O'odham Nation. The proliferation of geologic mysteries revealed by these studies compelled the crew to look eastward to the virtually unstudied Baboquivari Mountains. As Gordon Haxel described them, "The Baboquivari Mountains proved a sort of geologic Rosetta Stone—mapping of this range clarified or solved most of the mysteries posed by the ranges mapped earlier".

In this guide we hope to introduce the reader and visitor to some of the remarkable rocks of the area in and around Brown Canyon. We will also present some of the major concepts that have come out of the study of the range and explore how its story fits into and illuminates the geologic history of southern Arizona.

The first part of the guide introduces and uses some basic geologic principles and concepts as we reconstruct the geologic history of Brown Canyon and interpret the geologic evolution of southern Arizona and the American Southwest.

The last part is a road and trail log that guides the visitor through Brown Canyon. At each "Rock-Stop" it describes and interprets the rocks and landscapes seen at that location.

Although Brown Canyon is protected and has restricted entry, the Refuge policies include active visitor services with educational programs and activities. They offer regular guided hikes and special events in Brown Canyon. In addition, lodging is available within the canyon—an ideal way to explore the canyon. The Refuge staff can be reached at (520) 823-4251. Ask for the Assistant Manager or Visitor Services. The official website is at http://www.fws.gov/refuge/buenos_aires/.

We hope you enjoy this exploration of a rare and special area.

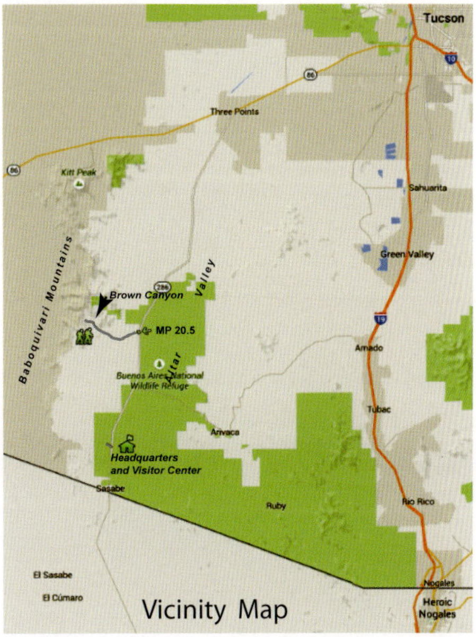

The Regional & Geologic Setting of the Baboquivari Mountains

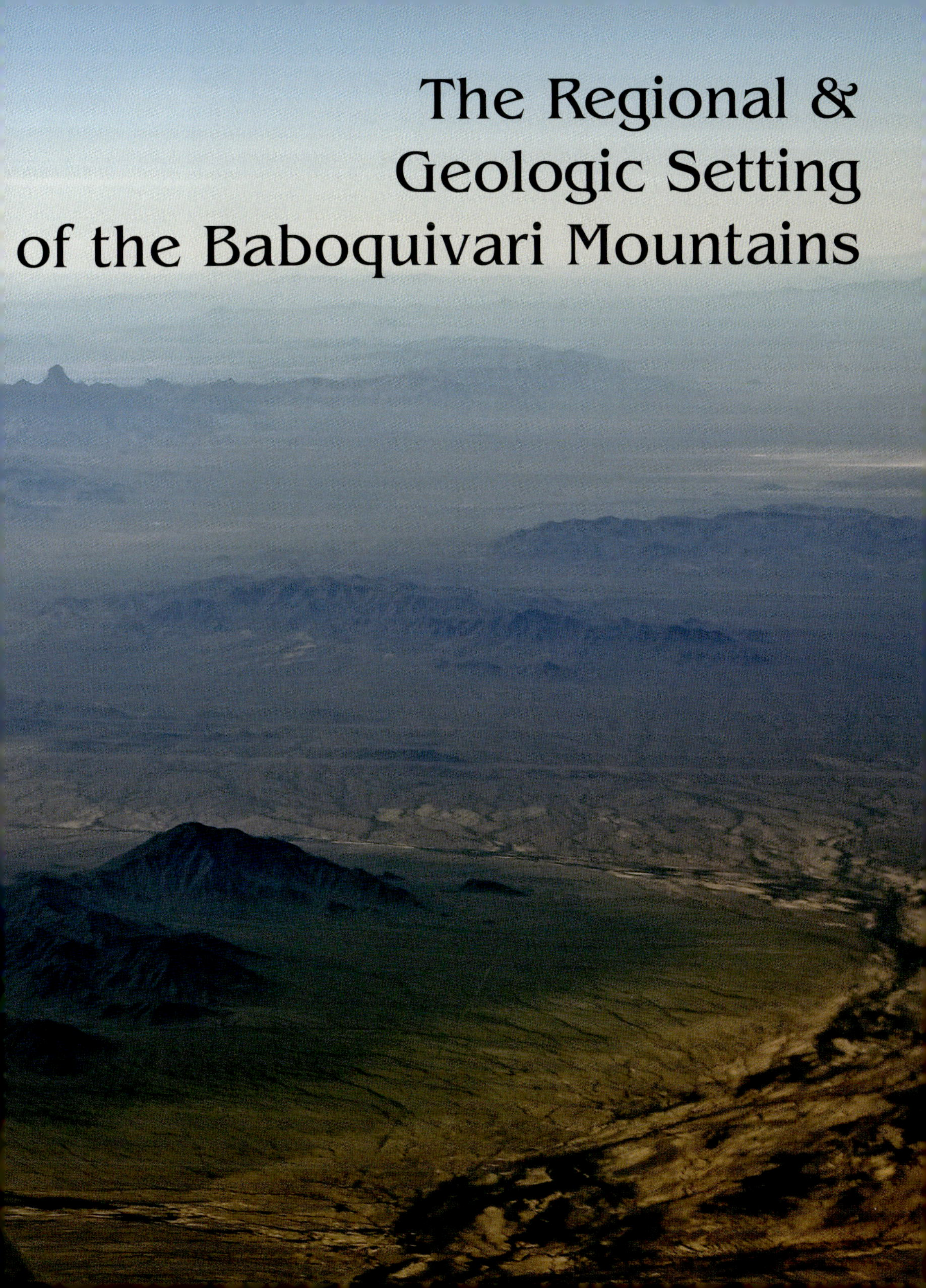

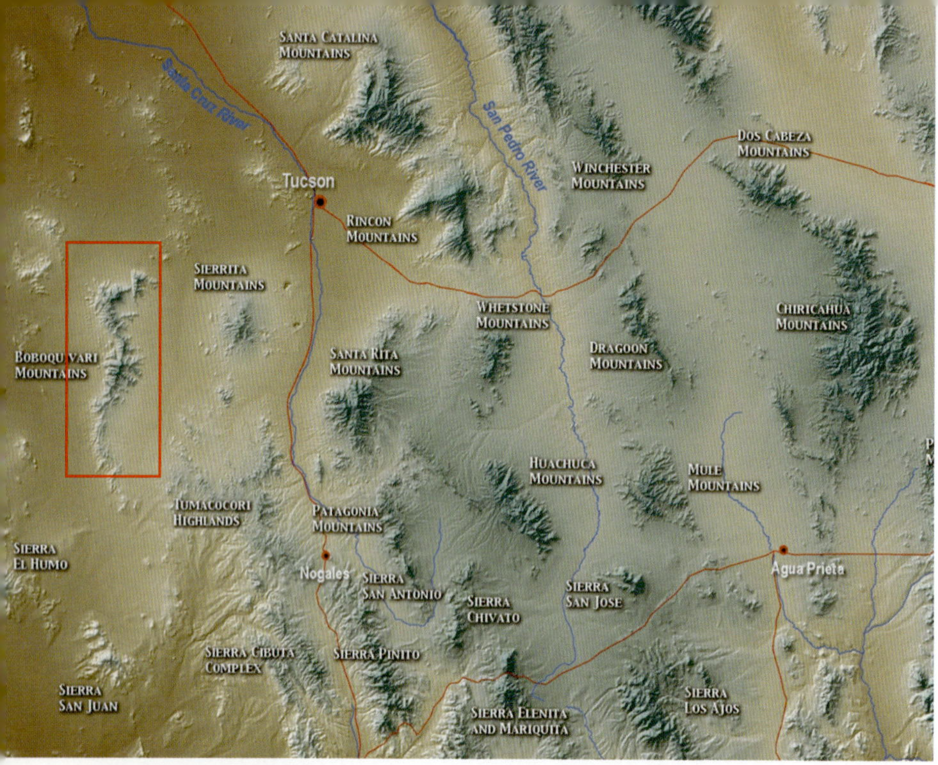

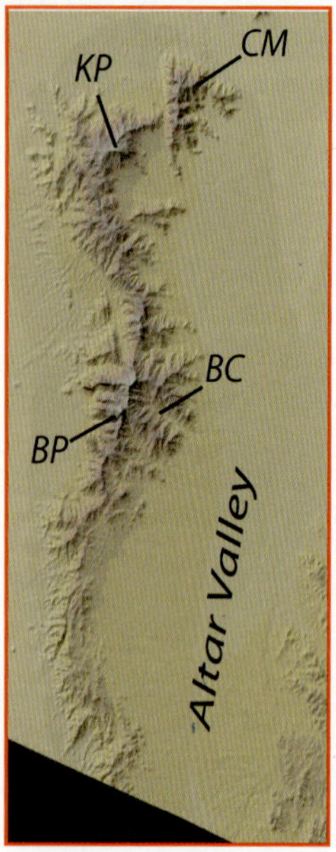

The Sky Islands

The Baboquivari Mountains are the western-most mountain range in a region of Arizona, New Mexico and Sonora, Mexico called the Madrean Sky Islands. The name alludes to the geographic isolation and distinct ecologic characteristics of the ranges. One of the factors in their ecologic isolation is the contrast in elevation between the basins and ranges. Baboquivari Peak at 7734 ft (3275 m) is in the middle of the range which extends 40 miles (64 km) from its south end in Sonora to Kitt Peak at 6877 ft (2199 m). The Altar Valley is about 3000 ft (900 m) elevation in the area east of Baboquivari Peak and the Baboquivari Valley is as low as about 2200 ft (670 m).

The entire range is rugged, wild, and contains areas little changed from their state before the arrival of Europeans. East of the range is the Altar Valley which contains 118,000 acres of desert grasslands protected by the Buenos Aires National Wildlife Refuge. Ranchers and other property owners in the valley have managed their lands so that the valley still exhibits some features

Top of page: the Madrean Sky Islands of southeast Arizona and Sonora. Left: the vicinity of the Baboquivari Mountains. Baboquivari Peak (BP), Kitt Peak (KP), Coyote Mountains (CM). See red outline boxes in regional maps above and on next page. Sky Island map, courtesy Sky Jacobs http://skyjacobs.com/, landforms image, USGS.

Previous pages: aerial view of the Baboquivari Mountains looking southeast, courtesy Malcom Andrews, http://aerialhorizon.photography/. The foreground area to the crest of the range is in the Tohono O'odham Nation. Mexico is at the horizon.

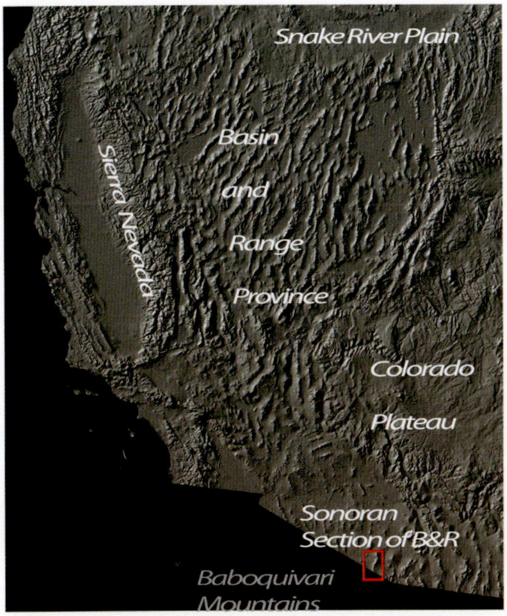

The Basin and Range and adjacent provinces. The landforms in Nevada were likened by early geologists to an "army of worms". The Baboquivari Mountains are in the Sonoran region. Red boxes in this and the previous maps enclose the range.

of the Sonoran grasslands.

To the west is Baboquivari Valley and the lands of the Tohono O'odham Nation. Beyond is a great desert studded with isolated lower elevation mountains composed of diverse igneous, metamorphic, and volcanic rocks.

The Sky Islands are part of a vast region of the west with a unique history. It is called the Basin and Range Province and it extends from southern Oregon and the Snake River Plains in Idaho through Nevada and eastern California southward deep into Mexico. Southern Arizona is included in the Sonoran Desert region of this province.

The landscapes of southern Arizona developed in just the last few million years, but the oldest recorded events are preserved in 1650 million year old rocks called the Pinal Schist. The oldest known rocks exposed in Arizona are in the Grand Canyon and are 1840 million years old.

The Baboquivari Mountains and the Altar Valley, seen here looking south from Kitt Peak, is an example of a Basin and Range landscape.

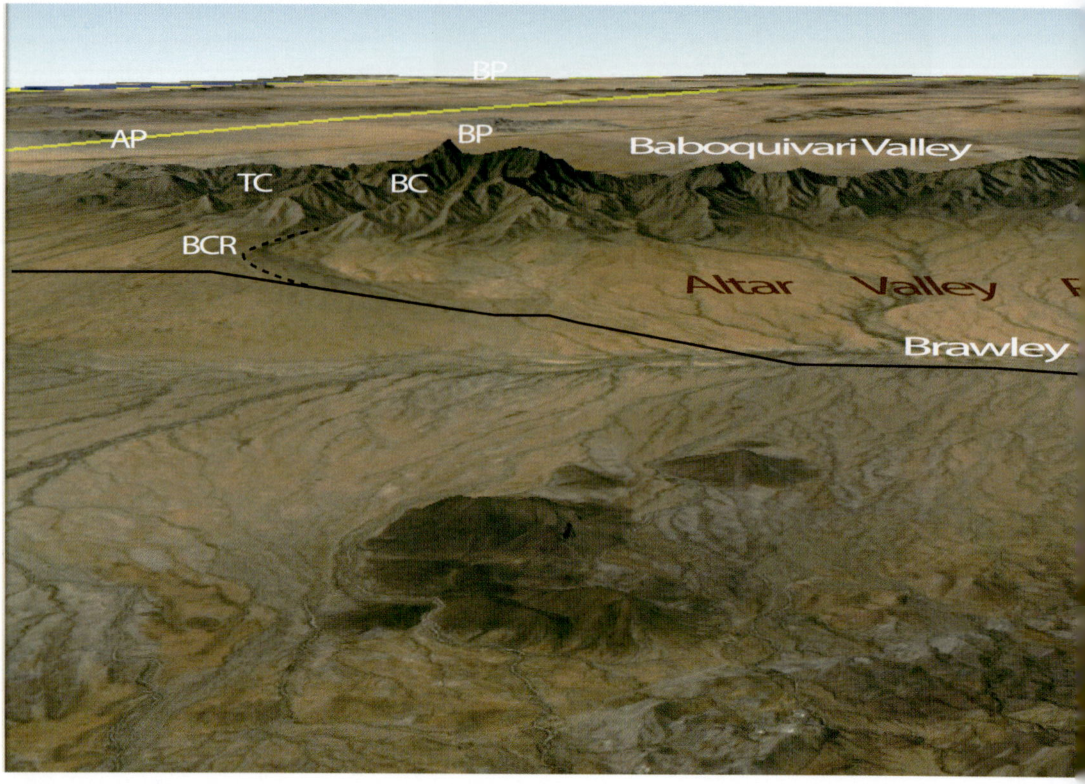

The Baboquivari Mountains and the Altar Valley: a Basin and Range Landscape

The Altar Valley, the Baboquivari Range, and the Baboquivari Valley capture the essentials of the Basin and Range Province. This west-looking view across the Altar Valley shows three main peaks. From north to south (right to left) these are the Coyote Mountains (CM), Kitt Peak (KP), and Baboquivari Peak (BP). South of Baboquivari Peak near the edge of the image is Aquirre Peak (AP).

South of Aguirre Peak, out of view to the south and astride the Mexican border are the Pozo Verde Mountains. In the middle of the image is the Altar Valley, with Brawley Wash, the main drainage for the valley. The head of Brown Canyon (BC) and the Brown Canyon road (BCR) can be seen below Baboquivari Peak (BP). Thomas Canyon (TC) is immediately south of Brown Canyon.

Streams draining eastward off of the Baboquivari Mountains cross a broad apron of alluvium before they join Brawley Wash. This surface is referred to as a piedmont or bajada and will be

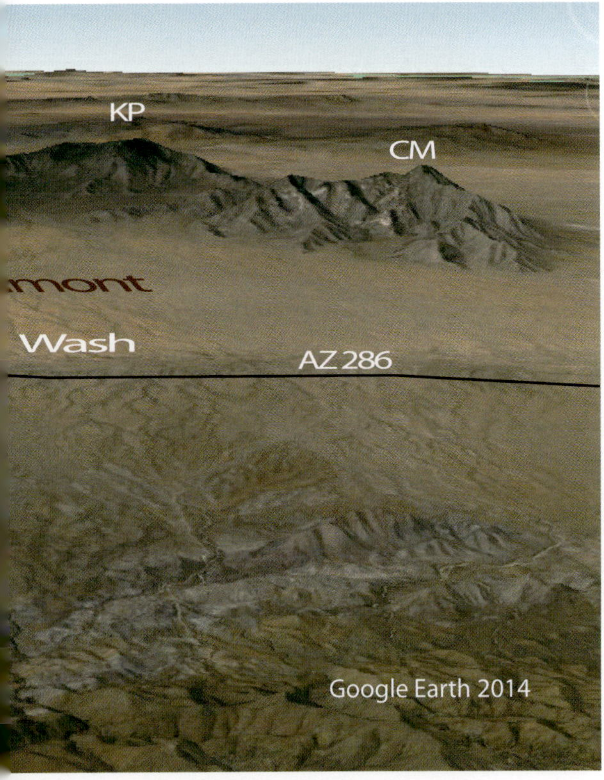

discussed in the chapter on landscapes.

In the right foreground are stripes of bedrock that are part of the Sierrita Mountains. The black areas are Paleogene basalt.

Beyond the Baboquivari Range is the Baboquivari Valley and beyond that is a terrain of smaller faulted ranges, all within the Tohono O'odham Nation. The yellow line in the upper left marks the border of Arizona and Sonora, Mexico.

Four Episodes in Arizona Geologic History

The geologic history of Arizona can be conveniently divided into four major episodes. The first episode was during the assembly of North America. The second episode was a time of marine flooding. The third was the beginning of great and prolonged volcanism. Much of this guide is devoted to this third episode because the Baboquivari Mountains have a tale of volcanoes to tell. The story continues through an episode of major continental deformation when the west was nearly rifted apart, and to more recent times when the modern landscapes were created. The standard geologic time scale immediately to the right shows ages in millions of years—**Ma**. The images and text summarize the significant events. When you study this display, be sure you start reading from the bottom with Episode I, and progress up through time to the latest episode and **NOW**.

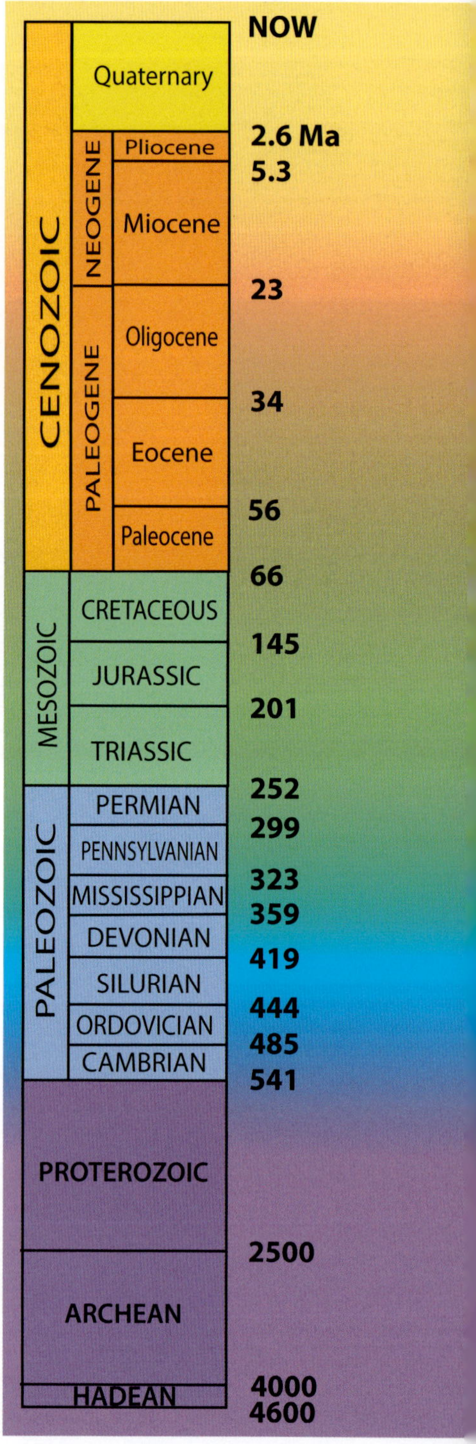

IV. After millions of years of collision and compression, North America began to rift apart. Much of the crust between the Rocky Mountains and the Sierra Nevada Mountains was stretched and broken into fault blocks. Some blocks slipped down to form basins and others stayed higher to make ranges. This is the Basin and Range Province and it is the foundation for the modern landscapes of southern Arizona. Photo courtesy Murray Bolesta, diagram from USGS.

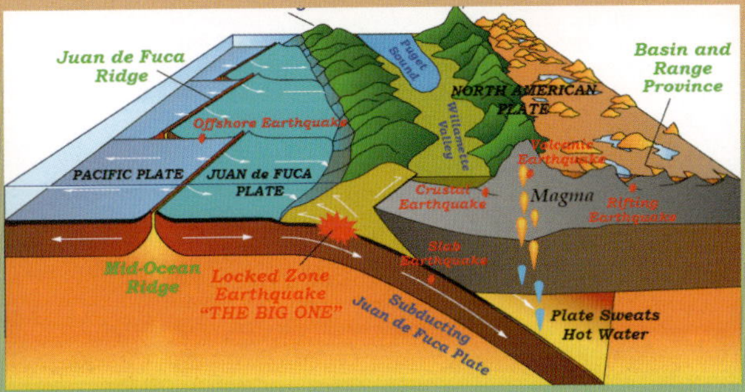

III. Beginning approximately 200 million years ago, North America collided with and overrode the crust of the Pacific Basin. This formed a subduction zone much like the Cascadian subduction zone of the Pacific Northwest today. This resulted in the eruption of a series of volcanoes across southern and western Arizona. Significant igneous activity continued through the Paleogene. Image courtesy Robert J. Lillie.

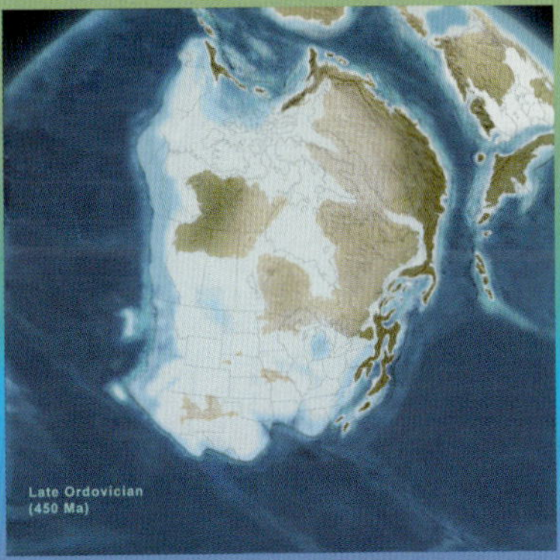

II. By 450 Ma, a continent had grown that would one day be called North America. During Paleozoic times much of the continent was flooded by a shallow sea. This image shows North America 370 million years before the story of the Baboquivari Mountains began. Most of the limestone found in Arizona was deposited in this sea over millions of years. Many of the marine fossils found in Arizona were deposited in this sea. © Ron Blakey, Colorado Plateau Geosystems.

I. Continents began to form about 2500 Ma. They grew, collided, rifted and fragmented as they wandered about the Earth. By the middle Proterozoic the ancient core of North America had formed. It is called Laurentia and is in the center of this globe. An important feature then and now are "island arcs" like Japan and Indonesia where volcanic material erupts from the sea. The image depicts many island arcs and crustal fragments that might collide to help form continents. © Ron Blakey, Colorado Plateau Geosystems.

The Rocks of Brown Canyon

Reading the Bedrock of Brown Canyon

The Earth's crust is made of rocks which are composed of minerals—solid, naturally occurring chemical substances. The distinction between rocks and minerals is an important one. Rocks are made of minerals and minerals are made of atoms.

Geologists recognize three fundamental types of rock—igneous, sedimentary, and metamorphic. The rock cycle diagram shows the three kinds of rock and the geological processes that link them. All three types occur in the Baboquivari Mountains, but the most important for our story are igneous rocks. These were formed by the cooling and crystallization of magma, which is molten rock. Both extrusive (volcanic) and intrusive (plutonic) rocks were created. We will discuss igneous rocks in detail in the next section.

Sedimentary rocks are formed by lithifying processes which change loose sediment into hard rock. The sediment was deposited by wind, water, ice, and gravity. Sedimentary rocks are typically layered like those in the Grand Canyon. Brown Canyon sedimentary rocks are not common, not typical, and are difficult to recognize.

The part of the crust exposed around Brown Canyon was deep and dominated by igneous processes. The contemporary surface rocks can be seen in other ranges.

Above: folded Jurassic siltstone in Brown Canyon. The original silt was deposited in a low-energy stream environment.

Aerial photograph on previous pages of Baboquivari Peak from the east. Divide in center separates Thomas Canyon on the left from Brown Canyon. Allison Camp Rhyolite dikes are in the foreground. Image courtesy Mary Scott.

Image courtesy of
The Canadian Museum of Nature.

Metamorphic rocks do not occur in Brown Canyon, but a few miles south, some of the rocks seen in Brown Canyon have undergone significant metamorphism. Metamorphic rocks also occur at the north end of the range in the Coyote Mountains. Metamorphic rocks testify to times of intense heating, chemical changes, stress, and deformation that transformed the original rock into a metamorphic rock.

Above: spectacular exposures of rocks at Pan Tak Pass between Kitt Peak and the Coyote Mountains. The rock, although igneous, exhibits a strong foliation (parallel alignment of structures) similar to that which occurs in many metamorphic rocks. These rocks were created deep in the crust under the conditions of intense deformation, heating, and invasion of magma; the same conditions that produce metamorphic rocks.

Igneous Rocks of Brown Canyon

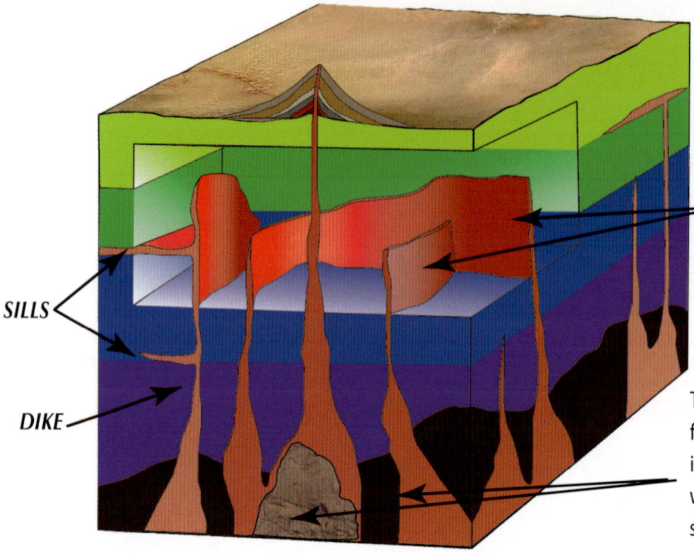

DIKES and SILLS are sheets of crystallized magma that invaded cracks and other weak zones in the crust. Dikes have formed the walls and towers on the west wall of Brown Canyon. See the top photo opposite.

STOCKS are larger plutons. The photo of Baboquivari Peak from the opposite page is inset into the diagram to show where the Baboquivari Peak stock might have formed.

A magmatic system can be very complex, including the magma source and all the plumbing and conditions that determine whether the magma ultimately rises to the surface and creates a volcanic center or stops short of the surface and becomes part of an intrusive mass that cools and crystallizes in deeper and hotter places.

The diagram above shows some features of a magmatic system. The host rocks, shown with black, purple, blue, and green colors in the diagram, have been intruded by magma that cooled and created plutons. The deeper larger plutons are called stocks and are connected to even deeper magma sources. The Baboquivari Peak pluton is a stock. Plutons can also form by filling cracks, faults, or pipes. In this way magma can rise to the surface to create volcanic centers. The cracks filled with crystallized magma can form dikes when the rocks are weathered and eroded.

The rocks of Baboquivari Peak are part of a stock of granite and the spires and walls at the back of Brown Canyon are rhyolite dikes. The rocks in these features crystallized from magmas that were rich in silica and formed light-colored rocks.

Granite is a light-colored rock that contains crystals large enough to be seen without magnification (coarse-grained texture). This means that the crystals had enough time to grow larger, conditions typical of larger deeper plutons, like stocks. In contrast, rhyolite is a light-colored rock characterized by very small crystals that are difficult to see without magnification (fine-grained texture). Such crystals were unable to grow large and are common in plutons that cooled quickly as volcanic rocks or at shallow depths referred to as subvolcanic or hypabyssal.

The west wall of Brown Canyon shows a spectacular display of rhyolite dikes. The high escarpment at the upper left is the face of a dike. Baboquivari Peak is just out of view to the left.

Baboquivari Peak is a stock of granite which cooled and crystallized several kilometers beneath the surface.

Intrusive contacts can be used to determine the relative age of features and interpret geologic history. The pattern of younger intrusive features that interrupt the pattern of older features is called cross-cutting relationships. The diagram to the left is a section from the Brown Canyon cross section on pages 23 and 24. It shows dikes in red, granite in orange, and sedimentary rocks in green. Notice how the pattern of the granite interrupts (cross-cuts) that of the sedimentary rocks. This means that the granite is younger than, and has intruded the sedimentary rocks. Now notice that the dikes cross-cut and are therefore younger than both. Faults can be analyzed in the same way. If a fault cross-cuts and offsets a feature, the fault must be younger than that feature.

The relative age of rocks tells us only whether a rock or feature is older or younger than another and differs from numerical or absolute age. The idea of relative age can be applied to sedimentary rocks. Since they are layered, the oldest layer is on the bottom and each successive layer is younger. In the diagram there are two sedimentary formations shown in addition to the igneous ones. The order of all events is (1) deposition of darker green formation, (2) deposition of lighter green formation, (3) intrusion of granite, and (4) intrusion of dikes.

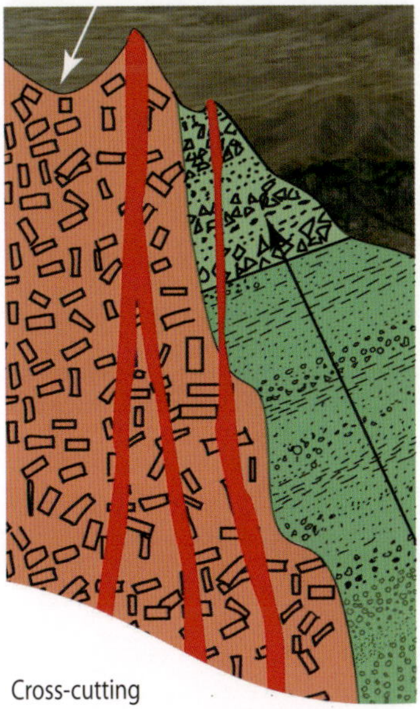

Cross-cutting relationships. See adjacent text.

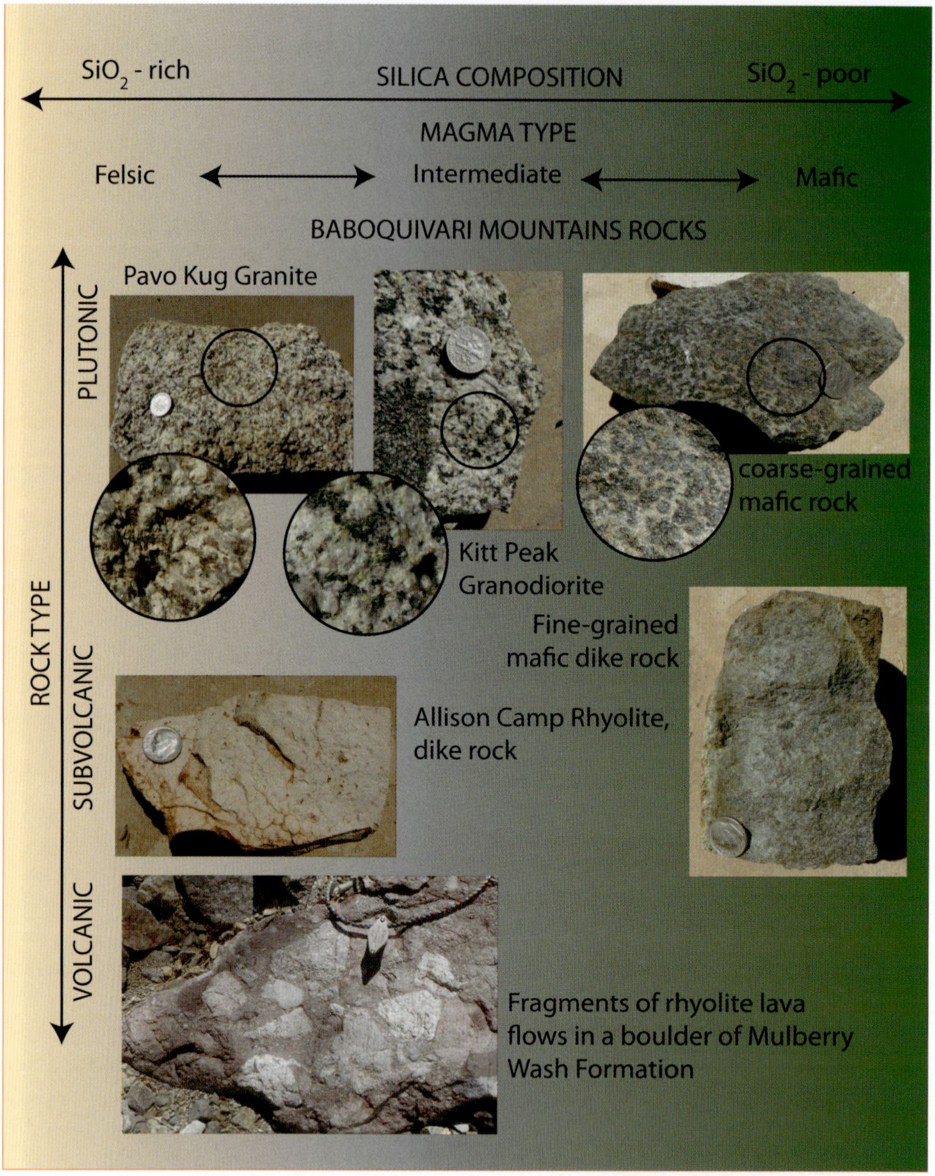
SiO2 - rich
SILICA COMPOSITION
SiO2 - poor
MAGMA TYPE
Felsic
Intermediate
Mafic
BABOQUIVARI MOUNTAINS ROCKS
ROCK TYPE
PLUTONIC
SUBVOLCANIC
VOLCANIC
Pavo Kug Granite
Kitt Peak Granodiorite
coarse-grained mafic rock
Fine-grained mafic dike rock
Allison Camp Rhyolite, dike rock
Fragments of rhyolite lava flows in a boulder of Mulberry Wash Formation

Two ways of organizing igneous rocks.

The diagrams on these two pages depict ways that geologists organize and name igneous rocks. The diagram on the previous page places common Baboquivari Mountains igneous rocks in a chart that relates rock type and magma type. The vertical axis shows how magma history is affected by rate of cooling. Slower cooling = PLUTONIC (coarser grain size), faster cooling = VOLCANIC (finer grain size). Subvolcanic means that the magma cooled faster near the surface in dikes or sills. In this simple view, plutonic rocks are coarser grained than volcanic rocks. The horizontal axis shows how silica (SiO_2) composition influences the appearance (mineral grains and color) of a rock. Silica-rich rocks are lighter colored (felsic) and silica-poor rocks are darker colored (mafic). The black arrows along the top and left side indicate that, in the real world, there is a continual gradation from one extreme to the other.

The diagram below plots mineral composition against silica composition with numerical values. This allows for a more precise method of classifying and naming. The top three rows of the diagram correspond to the entire diagram on the previous page without the pictures and local formation names. The mineral area shows fields demonstrating how the abundance of minerals relative to one another changes as the silica composition of the total rock changes. Two interesting observations that can be made are that there are just a few particular minerals that can crystallize out of a magma of a certain composition and that some minerals cannot coexist in equilibrium conditions. The critical factor is silica composition. Looking at the diagram, we can tell at a glance which are silica-rich minerals and which are silica-poor. The same can be said of the rocks.

To use the diagram accurately, we must be able to estimate the percentage mineral composition of the rock. As an example, consider a rock that has a composition corresponding to that of the red line and 67%. By using the scale on the vertical axis to measure the vertical width of each mineral field, this rock contains 13% biotite, 43% Na-plagioclase, 27% alkalai feldspar and 17% quartz. If this was a plutonic rock it would be called granodiorite, if it was a volcanic or subvolcanic rock, it would be dacite.

Obviously this method requires more information to use accurately, but it effectively demonstrates the concepts used and allows us to relate mineral composition to the rock, whether we can see the mineral grains or not.

Mineral Composition and the naming of Igneous Rocks

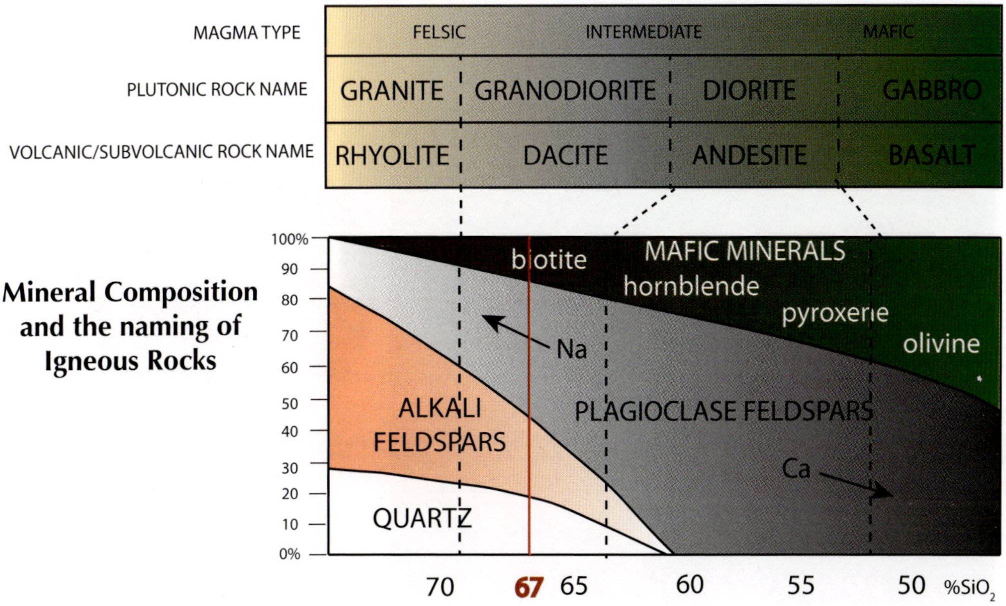

Two examples of porphyritic texture.

The rock above shows large feldspar crystals imbedded in a groundmass of fine-grained minerals. Note the few mafic minerals of intermediate size. The large crystals are about 1.5 cm long. The rock is a rhyolite.

The rock below shows feldspar crystals imbedded in a much darker groundmass composed of mafic minerals and feldspar. The rock is an andesite that occurs near Arivaca.

A rock with porphyritic texture exhibits significantly larger crystals imbedded in a groundmass of smaller grains. This is where the name porphyry comes from. Southern Arizona has world-class copper deposits and many important ore-bearing rocks have a porphyritic texture. These ore bodies are frequently called porphyry copper deposits.

Many of the dikes in the Baboquivari Mountains are spectacularly porphyritic. At the left is an example that shows prominent crystals of feldspar in a matrix of smaller mineral crystals.

Interpreting the porphyritic texture is a great example of reading history in the rocks. A porphyry is interpreted as having an early cooling period when the big crystals had a chance to grow, followed by a later episode where the remaining magma cooled quickly, imbedding the older crystals in a mass of younger and much smaller crystals. This can happen as a magma, a mush of crystals and liquid magma, is moved from a hotter place to a cooler, shallower zone or even to the surface in a volcanic eruption.

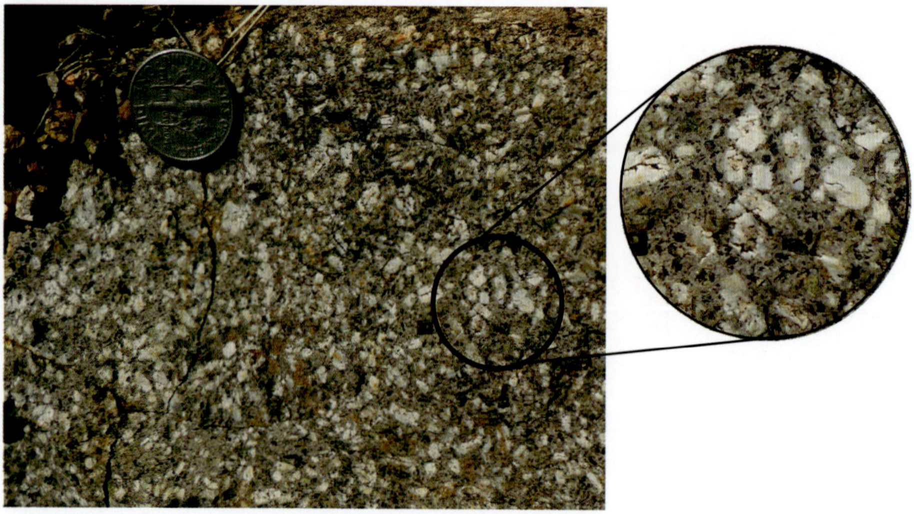

Rock-Forming Minerals in Brown Canyon

About 4900 mineral species are recognized. Over 90 percent of these are silicates, minerals that contain the elements silicon and oxygen. Silicates include most of the rock-forming minerals and, fortunately, there are only a few of these required for understanding the rocks around Brown Canyon.

A close-up view of a plutonic igneous rock from Kitt Peak called granodiorite. The photo shows several minerals that can be identified in Brown Canyon igneous rocks and used to interpret the history of the rocks. The quartz occurs as small shapeless translucent grey masses. The feldspar shows a roughly rectangular crystal shape, is opaque, and flashes when viewed from the proper angle because it has broken along a precise cleavage direction. The hornblende occurs as angular masses of black crystals. Some individual crystals show its rectangular crystal habit.

QUARTZ is silica (SiO_2), the fundamental chemical of magma and most rock-forming minerals. A magma does not crystallize quartz unless it has abundant silica. In a rock that has crystallized quartz from a mush of cooling magma along with other mineral crystals, quartz appears as shapeless grey translucent glittery grains.

FELDSPARS are the most common minerals in the Earth's crust. They are made of silicon, oxygen, potassium, sodium, calcium and aluminum. In a rock, feldspar is normally light-colored, more likely to show geometric crystal shapes than quartz, and frequently shows flashing cleavage faces which distinguishes it from quartz. Along with quartz they are the most important minerals of granite.

Two groups of feldspar are distinquished chemically. The alkali feldspars, named for the alkali elements sodium (Na) and potassium (K), vary continuously between $NaAlSi_3O_8$ (albite) and $KAlSi_3O_8$ (orthoclase). The plagioclase feldspars vary from $NaAlSi_3O_8$ to $CaAl_2Si_2O_8$ (anorthite).

A distinctive form of alkali feldspar found in the Baboquivari Peak Perthite Granite is called perthite. It is composed of very thin intergrown lamallae of albite and orthoclase that can give it a striking luster and patterning. Some perthite is considered to be a semi-precious gem.

Above: Baboquivari Peak Perthite Granite from Brown Canyon. The flesh-colored mineral is perthite. The original mafic minerals have been altered to hematite and chlorite.

MAFIC MINERALS are dark-colored because they are relatively poor in silica and rich in iron and magnesium. The most common mafic minerals in Brown Canyon rocks are hornblende and biotite.

EPIDOTE is a pistachio green mineral frequently seen in Brown Canyon. It is a calcium aluminum iron silicate and is the result of the chemical alteration of earlier minerals. It can be seen as veins, pockets of coloration, and as replacement of feldpar crystals.

Right above: clusters of hornblende and biotite in a granodiorite. The light grey grains are quartz and the opaque white mineral is feldspar.

Right below: a feldspar crystal about 2 cm long replaced by epidote.

How Do We Know the Ages of the Rocks?
Zircon as Nature's Little Timekeeper
by Gordon Haxel, PhD

Four major igneous rock units of the Brown Canyon area—Ali Molina Formation, Pavo Kug Granite, Baboquivari Peak Perthite Granite, and Allison Camp Rhyolite—are respectively 170, 159, 146, and 26 million years old (each ± 2–4 Myr). These Jurassic and Tertiary ages are far older than recorded human history, measured in tens of thousands of years, but far younger than the age of the Earth and Solar System, about 4.6 billion years.

How do geologists determine the numerical ages of ancient rocks? Several methods are used, but the most reliable and versatile involves the mineral zircon, which forms in trace quantities during the crystallization of many common igneous rocks. Zircon ($ZrSiO_4$) is composed of three essential chemical elements—zirconium, silicon, and oxygen. If it contained only these elements, zircon would not be useful as a timekeeper. However, when zircon crystallizes it usually incorporates a small amount of uranium, a radioactive element. Slowly, over geologic time, uranium decays to lead, at a well-known and absolutely constant rate. Zircon thus provides a geologic clock. By measuring the concentrations of uranium and lead in zircon, geologists can calculate the time elapsed since the zircon, and therefore the igneous rock containing the zircon, crystallized. This date is called the uranium-lead age, expressed in millions or billions of years before present.

Photomicrograph of zircon crystals about 0.5 mm long. Image, National Science Foundation.

But, the story gets better. Natural uranium comprises two varieties (isotopes), U-235 and U-238. Each of these two isotopes of uranium decays to a different isotope of lead. The two decay schemes are independent of one another and proceed at markedly different rates. By precisely measuring concentrations of both isotopes of uranium and several isotopes of lead in the same zircon crystal or sample, geologists can calculate two independent ages. If these two uranium-lead ages agree (are concordant), that date is considered the age of the zircon and the rock. If the two ages disagree (are discordant), the situation is more complicated but not necessarily hopeless. The most common causes of discordant ages are partial loss of lead from zircon, yielding apparent dates that are too young; or inheritance of zircon or lead from older rocks, giving apparent dates too old. Often, by determining apparent ages for multiple crystals or samples of zircon, corrections can be calculated and a unique, robust crystallization age obtained. Uranium-lead ages can also be determined for other minerals in addition to or instead of zircon, or analogous dating methods that exploit other long-lived radioactive elements (such as potassium, rubidium, samarium, lutetium, rhenium, or thorium) can be used. Most igneous rocks are dateable by some combination of these techniques.

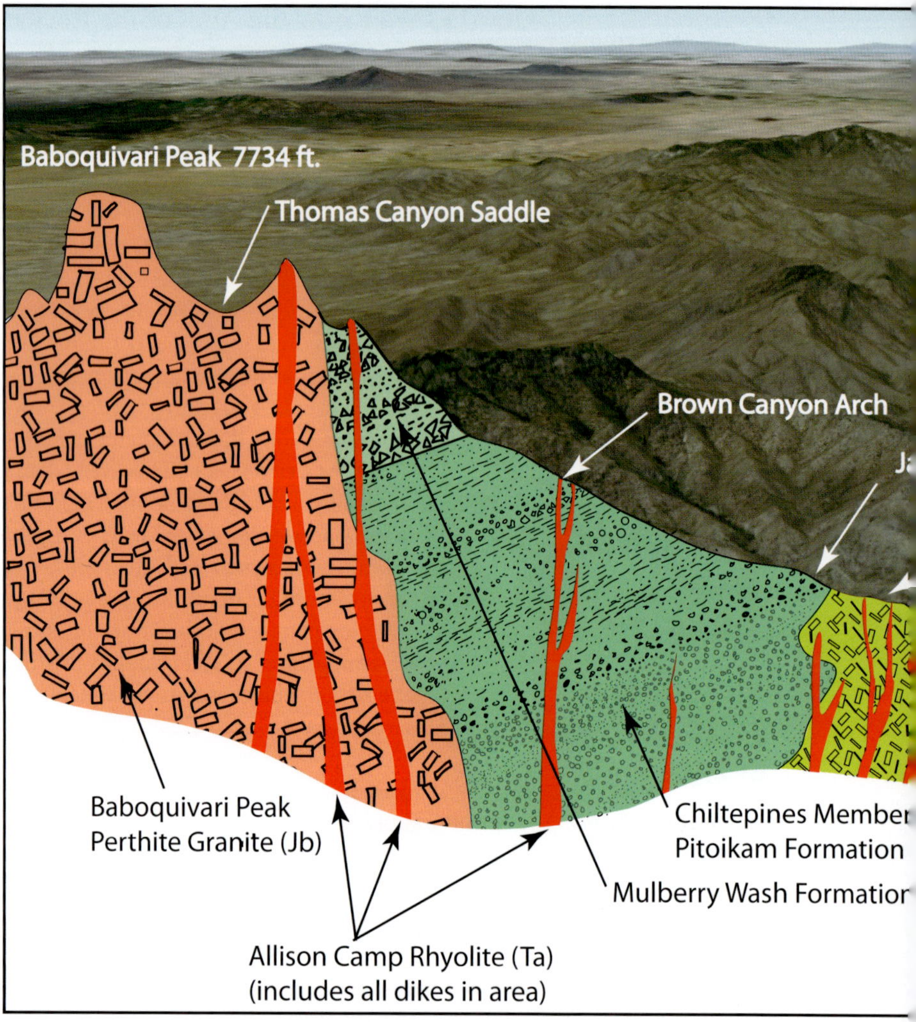

Geologic Illustrations for Brown Canyon

Geologists rely heavily on imagery to communicate their ideas. Two images that are very common are the geologic map and the geologic cross section. The Brown Canyon geologic map is at the back of the book and the cross section is above. A geologic map shows the areal distribution of different rocks exposed at the surface or inferred to be just beneath a covering of recent loose sediment. The map uses different colors and/or patterns to show areas that represent the rocks or other earth materials of interest. Each has a corresponding text symbol that identifies it uniquely. As an example, look at the map and find the colored area with the symbol "Jt". This is called the Tinaja Springs Porphyry. The J stands for Jurassic and the "t" for the formation name.

A geological formation is a recognizable and mappable set of associated rocks such as

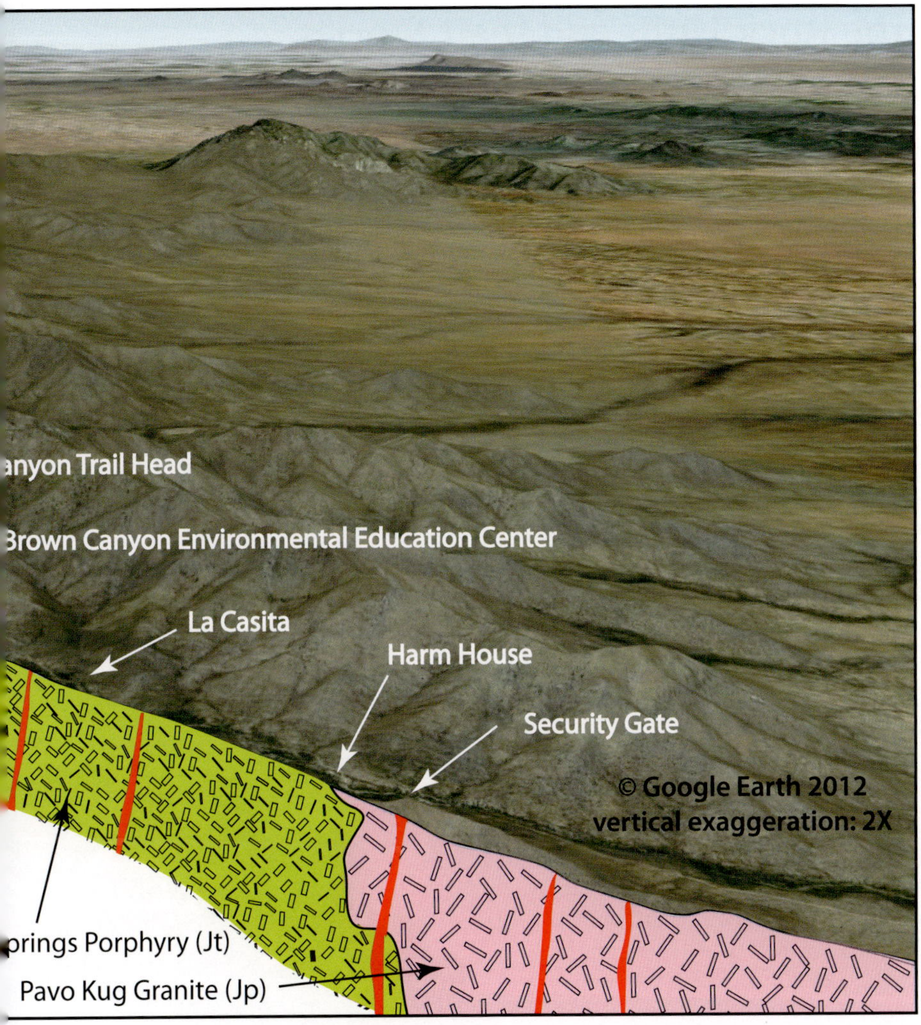

granite, volcanic rocks, or sedimentary rocks. Usually these are formally defined geologic units that have official names acknowledged by the profession. To learn more about the formations, study the explanation that accompanies the geologic map.

Other important features mapped are faults, dikes, strike and dip (the inclination of bedding), and the contacts between two adjacent formations.

The geologic cross section above is a vertical view or slice through the area and shows an edge-view of the rocks. The slice corresponds approximately to the line labeled A-B on the map. Lithologic symbols have been added to the cross section to give you a visual sense of the rock type.

Because the subsurface rock cannot be directly observed, cross sections are interpretive .

The Rock Units and their Stories

A geologically informed observer has a special ability to see beyond what they are looking at. A landscape viewed today becomes just the latest slice of a continuously changing series of landscapes or seascapes and the geological observer can create and visualize an ancient landscape. The same can be said of looking at a rock, discovering the historical clues, and inferring the history of that rock. The art and science of "reading the rocks" is one of gathering and interpreting the clues and logically recreating, from the evidence, the story of a rock, a landscape, and a piece of our planet.

Geologists organize and formalize rocks into units called formations. Formations can be subdivided into members and collected into groups. Formations become the basis for analysis and conclusions about the nature and history of the rocks and the place they occupy in the crust. In the next pages are descriptions of each unit found in the Brown Canyon area and an interpretation of the rocks' origin and history. Primary to this is the place where the rock formed, whether it was magma cooling deep in the crust as a granite or a sediment being deposited in a particular setting such as a river gravel bar. The descriptions of rocks that originated on the Earth's surface include a depositional environment. Detailed studies of a rock allow geologists to associate a rock's place of origin with a modern environment. This information helps a geologist build a geographic map for that time. The paleogeographic maps we will look at depend on such information.

THE ALI MOLINA FORMATION

We begin with a formation that is not even seen in Brown Canyon, but very near by, because it sets the stage for our history and tells its own fascinating story of distant deserts to the north. The Ali Molina Formation is mostly rhyolite, an important volcanic and subvolcanic product of subduction zone volcanoes. It occurs as lava flows, flow breccias and tuffs, suggesting a volcanic vent or vents closeby. The Ali Molina Formation also contains beds

A composite image depicting what a southern Arizona volcano might have looked like during Ali Molina times when it was surrounded by dunes blown south from the area of Utah and northern Arizona. Terrain image of Namibia, Google Earth.

of sandstone that are clearly eolian (deposited as sand dunes).

Southern Arizona 170 million years ago was occupied by a magmatic arc. The landscapes were extensively volcanic and southeast Arizona may have looked like the Andes Mountains in Bolivia and Chile today (pages 31 and 32). Large explosive volcanoes were erupting lavas, plumes of ash, and avalanches of incandescent particles and hot gas. Rains gathered into streams which carried the volcanic material downstream to be deposited as sedimentary beds rich in volcanic material (volcaniclastic).

At the same time that volcanoes were erupting in southern Arizona, great sand-seas covered much of the area that is now the Colorado Plateau. These immense tracts of dunes, called ergs from the Arabic word, are now seen preserved in the spectacular redrock landscapes of northern Arizona and southern Utah. Winds carried strands of sand all the way to southern Arizona where the sand mingled with the volcanic deposits. It is easy to imagine the blowing sand covering volcanic rocks during times when the volcanoes were quiet, only to be buried with ash or lava during the next eruptive cycle. Similar volcanic rocks containing eolian deposits are also seen in the Santa Rita Mountains and in the area around Arivaca and the Pajarito highlands west of Rio Rico.

THE CHILTEPINES MEMBER OF THE PITOIKAM FORMATION

The Chiltepines rocks would be considered by most visitors to be uninteresting. They are not well exposed in Brown Canyon and the clues to their history are subtle. They are found in the upper part of the canyon where they are generally grey, fine-grained siltstones and sandstones with local pebble conglomerates. The Chiltepines Member is part of the mostly sedimentary Pitoikam Formation.

The Pitoikam Formation sediments are inferred to be the basin-fill of a large fault basin. The erosion of the uplands washed sediments across the border fault into the basin and deposited coarse-grained sandstones and conglomerates along the margins of the basin in a series of alluvial fans. This occurred during a non-volcanic interval. The Pitoikam basin might have resembled the feature labeled "B" on the map on page 35.

The finer grained sandstones and siltstones of the Chiltepines Member were deposited by a river flowing through the central part of the same basin.

Siltstone beds in the Chiltepines Member of the Pitoikam Formation. Lighter-colored layers are zones where chemical cement concentrated when the sediment was becoming a rock.

Beds of conglomerate indicate episodes of faster water. The sediments do not contain coarse local volcanic material so they must have come from a non-volcanic area in a river system that passed through the basin, possibly from the north. The rocks are different, but the terrain was not unlike that of today with erosion of the range filling a faulted basin and a through stream, like the Santa Cruz River, depositing sediments in its flood plain.

THE MULBERRY WASH FORMATION

By critical acclaim these are the most attractive rocks in Brown Canyon. They do not crop out in the lower canyons, but form part of the imposing west wall of the canyon. Fortunately, erosion has littered the canyon with cobbles and boulders composed of multicolored clasts of many sizes and shapes. Angular clasts indicate minimal geologic work. More rounded clasts indicate that they were picked up and transported by running water, gradually rounding them more and more as they traveled downstream. Geologists call a rock composed of predominately rounded fragments a conglomerate. The word breccia is used to describe a rock that is composed of angular clasts.

All the clasts are rhyolite and were originally deposited as lava flows, breccia flows, and landslide breccias with no evidence of sorting and layering. Commonly the broken volcanic rocks were picked up and transported by streams.

These rocks came from a volcano that was located a few miles to the north. The volcano's silica-rich, gas-charged magma produced vigorous eruptions that erupted lava flows and explosively ejected flow breccias. Pyroclastic flows spread ash and debris over many square miles. What the visitor to Brown Canyon sees in the Mulberry Wash rocks is the breccia and volcaniclastic sediments deposited by streams flowing down the slopes of the volcano.

A boulder of volcaniclastic Mulberry Wash Formation. The rhyolite clasts vary from angular to rounded, reflecting the amount of wear thay experienced during transport by streams.

Up to this time we have been telling a story of volcanoes and rivers because most of the historic clues came from Jurassic volcanic and sedimentary rocks exposed in the Baboquivari Mountains. Volcanism continued with vigor in the region, but the later record of surface processes recorded by volcanic and sedimentary rocks was not preserved in the Baboquivari Mountains. What is seen is the deeper part of the magmatic systems—plutonic rocks. These are the rocks that form the striking linear features that criss-cross the range (see the title photo), the spires and towers at the head of Brown Canyon, and Baboquivari Peak. Plutonic rocks also make up most of Kitt Peak and the Coyote Mountains to the north.

Recall that magmatism started in southern Arizona about 170 million years ago with the initiation of the Jurassic magmatic arc. Arc magmatism ended about 165 million years ago as compression diminished and the dominant forces became more extensional. Succeeding episodes of igneous activity reflected the complex Cretaceous and Neogene tectonic history of the area as it was affected by the Laramide Orogeny (70 Ma - 50 Ma) and the Basin and Range extension (30 Ma - 5 Ma). The rocks of the Baboquivari Mountains provide an impressive plutonic record of these times and Brown Canyon provides impressive displays.

TINAJA SPRINGS PORPHYRY

The central part of Brown Canyon and the flanking hillsides are underlain by the Tinaja Springs Porphyry. The Environmental Education Center is built on it and, in part, of it. The rock has an overall rusty buff color. The rock faintly glitters with tiny crystal faces. Under magnification it is seen to be composed of crystals of quartz and feldspar imbedded in a finer grained groundmass that varies in color depending on the amount of iron oxide present.

Although this rock is compositionally a rhyolite, it is also called a porphyry because it contains larger crystals imbedded in a groundmass of much smaller, commonly microscopic crystals. This is a typical feature of many igneous rocks in southern Arizona and is often associated with major copper deposits. In the mining world, the word "porphyry" stirs thoughts of copper riches, but not in Brown Canyon.

The Ali Molina and the Mulberry Wash Formations are composed of rhyolite and are very similar in composition to one another and to the Tinaja Springs Porphyry. Both the Mulberry Wash and the Ali Molina Formations are intruded by the Tinaja Springs Porphyry and came from the same magma source.

THE TOWERS AND WALLS OF BROWN CANYON

From any vantage point near Brown Canyon, a remarkable grid of criss-crossing walls and cliffs is visible running along hillsides or up a slope and across a ridge. A single wall forms the crest of the ridge that divides Brown Canyon from Thomas Canyon. As you travel up the canyon road, many craggy broken outcrops and walls descend the slope, cross the canyon bottom as bold outcrops, and ascend the opposite side. From the Environmental Education Center you can see the headwall with its extravagant display of great walls, buttresses, towers, and cockscombs (photo right). Rising behind the ridge, Baboquivari Peak oversees it all. As we look up at the headwall we can see evidence for three different magmatic episodes separated by millions of years. First is the Mulberry Wash Formation of 170 million years ago, then the Baboquivari Peak pluton of 146 million years ago, and then the much younger dikes of Allison Camp Rhyolite of 26 million years ago.

The dike rocks are commonly very pale-grey to yellowish-grey microcrystalline rhyolites with very low content of mafic minerals. The rock weathers to a dark color which disguises its true nature. Another important occurrence is a very porphyritic rhyolite containing perfectly shaped rectangular crystals of alkali feldspar imbedded in a groundmass of very fine-grained feldspar, quartz, and mafic minerals. The dikes' composition makes them resistant to weathering and they occur as isolated fractured outcrops surrounded by vegetation-covered slopes.

The southwest part of upper Brown Canyon: Baboquivari Peak, composed of perthite granite, is on the left behind the ridge separating Brown Canyon from Thomas Canyon. The towers, walls, and buttresses are dikes of the Allison Camp Rhyolite.

Brown Canyon and the Window into

Baboquivari Mountains: a Jurassic World

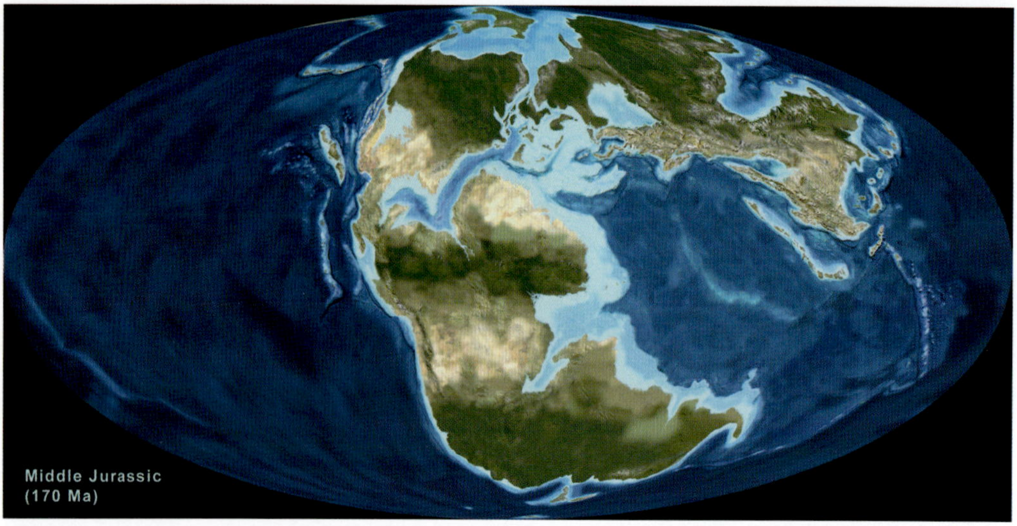

World paleogeography during Middle Jurassic time, 170 Ma. © Ron Blakey, Colorado Plateau Geosystems.

Jurassic Paleogeographies and Tectonics

The world during the time of the earliest Brown Canyon rocks was very different from today. About 250 million years ago the supercontinent called Pangea was assembled. The map above shows the Earth when the Atlantic Ocean was opening and driving North America westward toward the Pacific Ocean basin.

The map opposite shows North America with its west coast further east than today and offshore island arcs on a collision course with the continent. The accretion of these arcs added much of the land that is now part of California, Nevada, Oregon, and Washington.

The record that is preserved in the rocks of the Baboquivari Mountains begins at this time. Earlier geologic events are not represented in the Baboquivari Mountains. The record of these times does not occur here, or is buried beneath younger rocks.

The plate tectonics of the Jurassic Southwest included a collisional subduction zone boundary between North America and the Pacific Basin plates similar to the modern Cascadian subduction zone of Washington and Oregon (see diagram on page 36).

Previous pages: an aerial view southeast across Lauca National Park, Chile to Sajama National Park, Bolivia. This is the central magmatic arc of the Andes Mountains. Near the left margin of the photo is the extinct stratovolcano, Nevado Sajama, Bolivia's highest mountain. The crest of the Cordillera Occidental follows the line of volcanoes to the right. Most of the volcanoes are Pleistocene, but Parinacota (center of left page) has erupted in the last few thousand years. If this was the Jurassic arc of Arizona, the Baboquivari rocks would have beeen deposited in a position similar to that below the camera. Image © Flickr user Marcus Brutus.

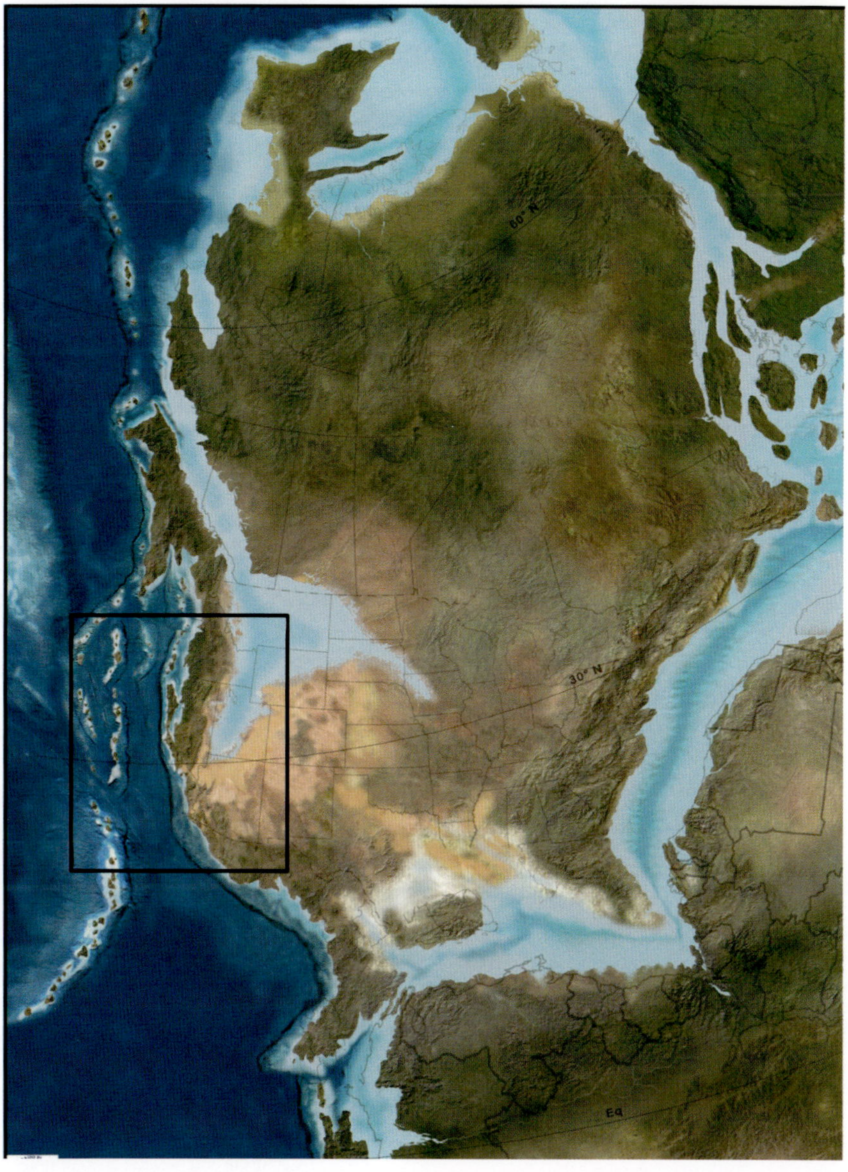

North America during Middle Jurassic time (170 Ma). The supercontinent Pangea is breaking up and Africa is pulling away from southeastern North America in the lower right. Europe is moving east in the upper right. Western North America is occupied by an inland sea, fringed on the south and east by desert sand-seas (ergs). A subduction zone along most of the west coast marks the collision of Pacific Basin plates with the continent. The strands of islands to the west are volcanic island arcs bound for collisions with North America. Inset shows the location of the map on the next page. © Ron Blakey, Colorado Plateau Geosystems.

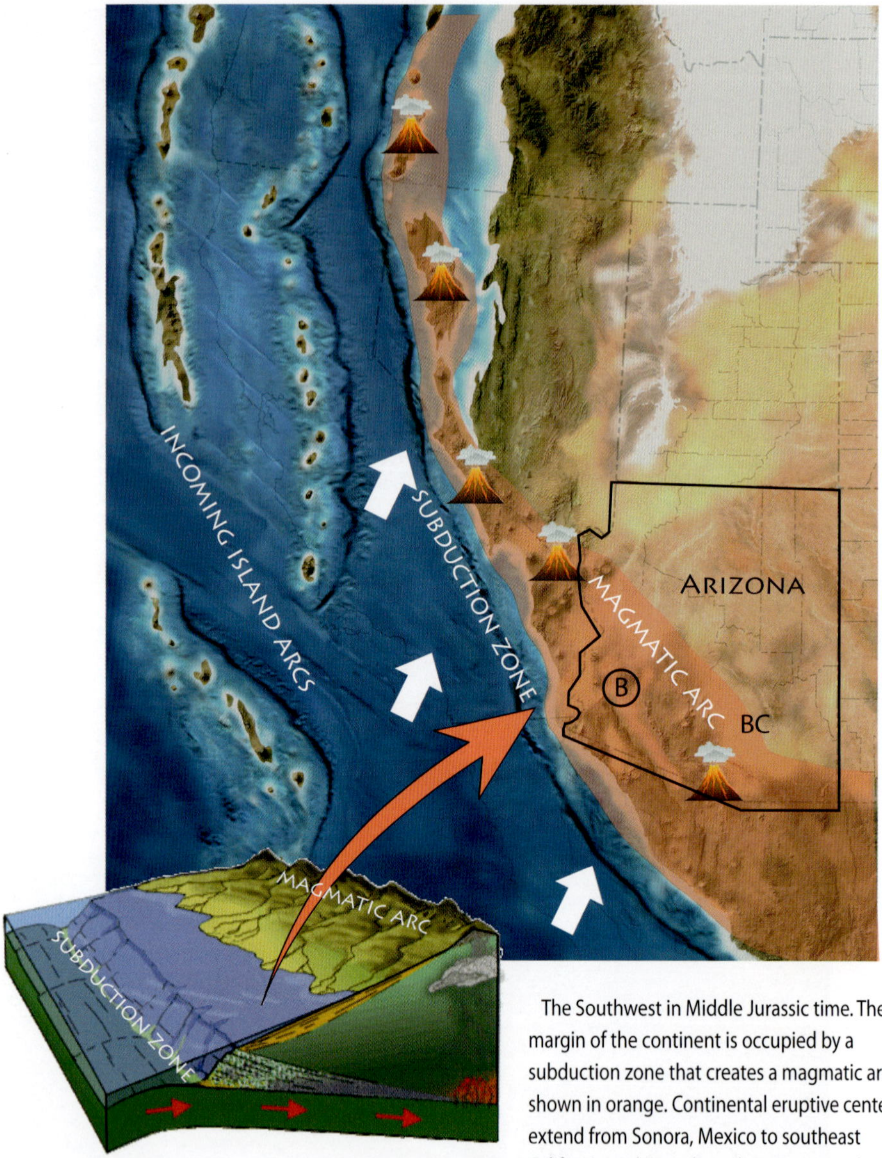

The Southwest in Middle Jurassic time. The margin of the continent is occupied by a subduction zone that creates a magmatic arc, shown in orange. Continental eruptive centers extend from Sonora, Mexico to southeast California and Nevada and continue northward as offshore islands. The inset diagram shows how a typical subduction zone could relate to the Jurassic Southwest. White arrows show the oblique angle at which the oceanic plate is moving towards the North American plate. Incoming islands will eventually become part of the North American continent. The present-day Baboquivari Mountains are near the eastern edge of the arc. The map also shows the desert that occurred in northern Arizona and southern Utah and played a role in the history of southern Arizona. The feature labeled "B" is referred to in the text description of the Pitoikam Formation on page 26. Paleogeographic map © Ron Blakey, Colorado Plateau Geosystems, subduction zone modified from an animation by UCSB Educational Multimedia Visualization Center, http://emvc.geol.ucsb.edu/ 2_infopgs/IP1GTect/cSubduction.html.

We usually think of plate collisions and subduction zones as involving head-on convergence that generates compression with reverse faults (see graphic at the bottom of this page). Naturally, things in nature are not so simple. Think of a collision where the colliding objects are also skidding past one another. The combination of shearing and compressive forces allows simultaneous transform boundary processes (skidding) and convergent boundary processes (collision).

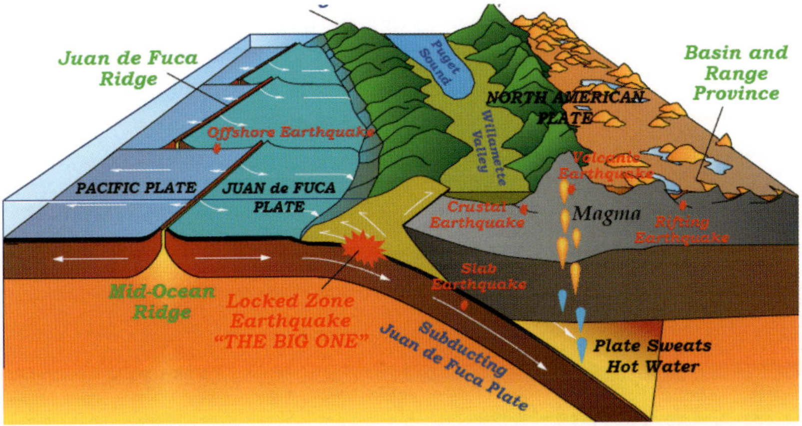

The Cascadian subduction zone is an example of a dominantly compressional convergent boundary. Magmatism builds a volcanic mountain range that is called a magmatic or volcanic arc. On the other hand, because of the oblique convergence, the Jurassic magmatic arc in Arizona had a combination of compressional and translational forces that allowed for extension (pulling apart) and basin formation within the main arc. This created fault blocks that collapsed along normal faults and formed basins. Such faults can extend deep in the crust and form passageways for magma to rise into the crust and feed volcanoes.

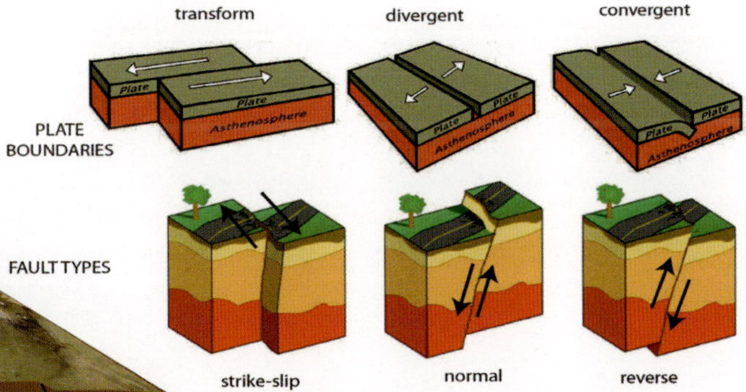

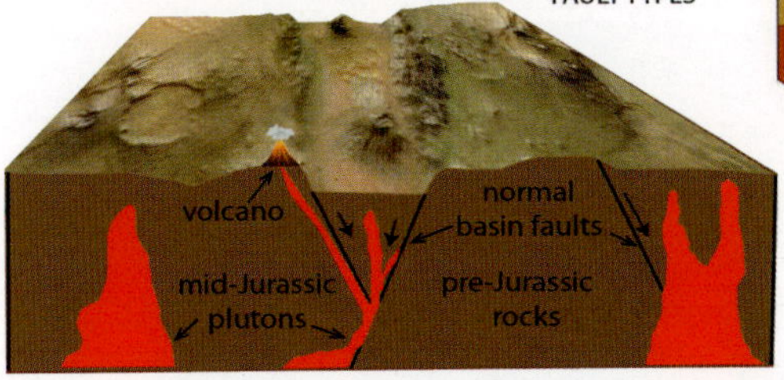

Top of page: the Cascadian subduction zone. Image courtesy of Robert J. Lillie, Oregon State University.

Lower right : the three types of plate boundaries and characteristic fault types. The upper row shows the relative motion of the interacting plates for each type of boundary with arrow pairs showing relative motion. The lower row illustrates the kinds of faults that are most common for each boundary type. Image, USGS.

Lower left: A block diagram depictiing what Jurassic basin B on the opposite page might have looked like. Note the invasion of magma, especialy that controled by normal faults. Adapted using terrain imagery created by Ron Blakey, Colorado Plateau Geosystems.

In Brown Canyon, the rocks of the Ali Molina Formation, the Mulberry Wash Formation, and the Tinaja Springs Porphyry record the main arc phase of magmatic activity. At Kitt Peak a suite of plutonic rocks also records this episode. By 160 million years ago the arc phase had ended and the crust in southern Arizona and Sonora, Mexico was starting to become even more extensional as translational strike-slip motion increased. New chambers of magma invaded the crust and began to supply a new phase of magmatism that used the extensional faults in transporting the magma into the crust. During this time the Pavo Kug Granite (159 Ma) was followed by the Baboquivari Peak Perthite Granite (146 Ma). From 146 million years ago until the intrusion of the Allison Camp Rhyolite dikes 26 million years ago there is no geologic record in the Brown Canyon area. To learn what was going on in southern Arizona during that interval of time and after, other places must be visited. Fortunately these are many, because Arizona has been very volcanic for a long time. Major episodes of magmatism occurred in the Cretaceous and the Paleogene, and have continued in different styles into recent times in local areas.

Brown Canyon's west wall. We can see three episodes of magmatic activity here. The oldest is the 170 Ma Mulberry Wash Formation volcanics. The next is the 146 Ma intrusion of the Baboquivari Peak Perthite Granite pluton, and the youngest is the 26 Ma Allison Camp Rhyolite dikes.

Southeast Arizona Landscapes

A landscape exists for only a moment in time. Anyone who has watched a flash flood knows this at the time scale of minutes. At the scale of tens to hundreds of years, processes we call erosion can cause significant changes, like the recent arroyo cutting in the Southwest. At the scale of hundreds of thousands of years, mountains and valleys will change dramatically, and at a scale of millions of years, mountain ranges can disappear, new ones can appear, and a continent can grow by hundreds of thousands of square miles—or break apart. So it should be no surprise that the scenery we enjoy has been around only a brief part of the continent's history. We have already learned of Andean-like mountains that preceded today's Sonoran desert country and we know that if we went back further it would have all been seabed!

Modern southern Arizona landscapes have emerged from a unique event that we have called the Basin and Range extension. From the beginning of our story in Middle Jurassic time until about 30 million years ago, western North America was the locus of a major collisional plate boundary with a subduction zone, terrane accretion, and dominant compressional deformation. After that the colliding oceanic plates of the Pacific Basin began to disappear beneath the North American plate and the transform boundary that would become the San Andreas Fault system began developing. Western North America was no longer in compression and began to expand westward. This had the effect of stretching the crust in the area that is now called the Basin and Range Province. The lower crust was ductile and stretched like taffy, but the upper crust was cooler, more brittle, and it broke into a mosaic of fault blocks that moved vertically and laterally relative to one

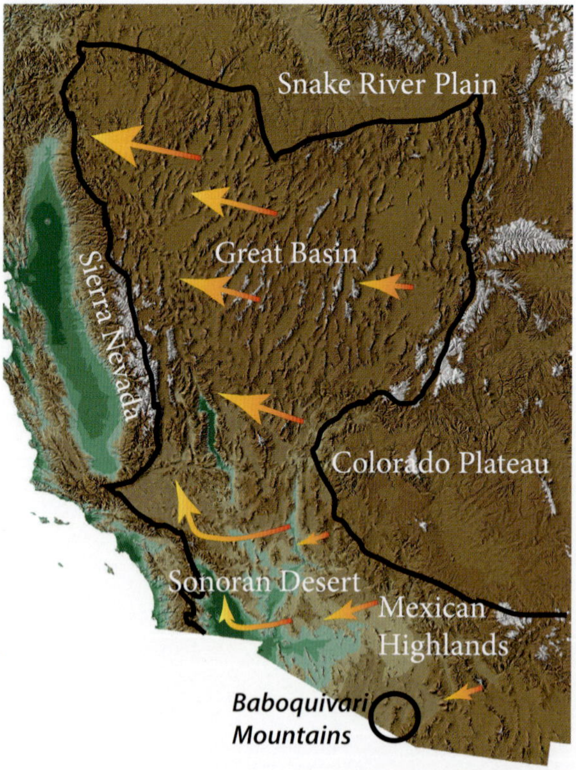

The Basin and Range Province and surrounding regions. The Basin and Range within this map area can be subdivided into the Great Basin, the Sonoran Desert and the Mexican Highlands. The Baboquvari Mountains are in the Sonoran Desert section. The orange arrows show direction and the approximate distance several ranges moved from 0 to 36 Ma. Curved lines show later motion influenced by northward translation of the Pacific plate (McQuarrie and Wernicke, 2005, *An animated tectonic reconstruction of southwestern North America since 36 Ma*). Map from USGS .

Previous two pages: a view eastward from the mouth of Brown Canyon. Beyond the Altar Valley, the near range is the Cerro Colorado Mountains and in the distance is the Santa Rita Mountains.

another. Some blocks remained relatively high, creating mountain ranges, while other blocks subsided along normal faults to form the basins.

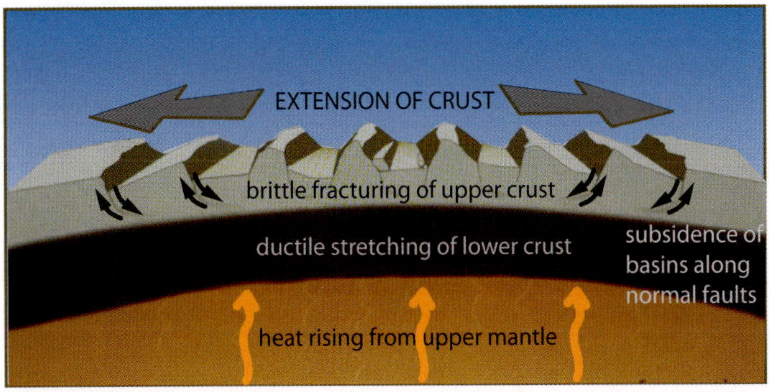

The tectonic processes that occurred during the extension of the western United States and Mexico from approximately 30 Ma to 5 Ma. Modified from an IRIS animation, http://www.iris.edu/hq/inclass/animation/basin__range_structural_evolution.

Through several millions of years and countless earthquake events, the basins subsided to form grabens as the rocks dropped along faults that bordered the basins and adjacent ranges. In a single large earthquake, the rocks may have moved several meters. After innumerable earthquakes the total movement added up to thousands of meters. At the same time, ranges moved apart from one another as the crust spread westward. As extension lessened about five million years ago in southeastern Arizona, earthquake events became less frequent. The last significant earthquake of this type occurred in 1887 in northern Sonora south of Douglas, Arizona. Geologists think that such major earthquakes are now happening every few thousand years in southeastern Arizona.

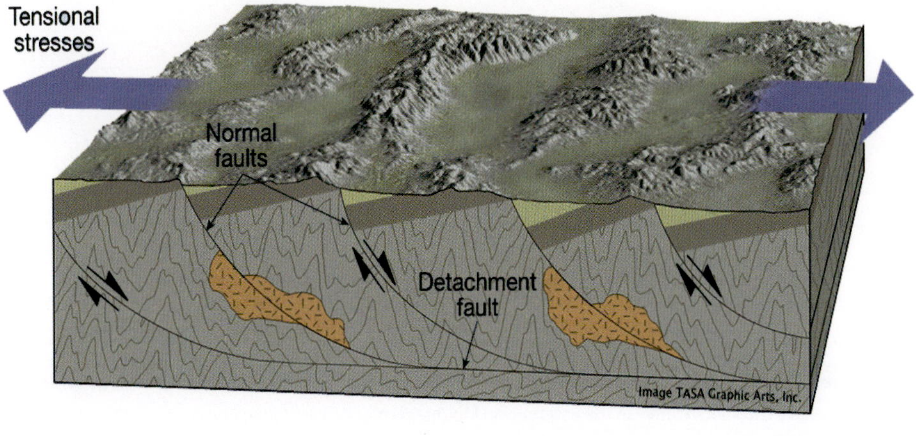

Block diagram showing how the upper crustal structure is expressed by the landforms. In this example, high angle normal faults flatten at depth and become part of a detachment zone during extension. Fault blocks above this zone subside to form half-graben basins and move horizontally under extensional forces. Image courtesy Tasa Graphics.

Filling of fault basins began as soon as topographic relief allowed for erosion by water and gravity. The basins subsided with each faulting event and water and sediment poured into the deepening basins, perhaps keeping pace with the subsidence. The water evaporated and the layers of gravel, sand, mud, and salts were left behind. Many closed basins with ephemeral playa lakes (salt flats) can be seen in Nevada. Willcox Basin, east of Tucson, is still closed and contains thousands of feet of sediment in addition to economic deposits of salt.

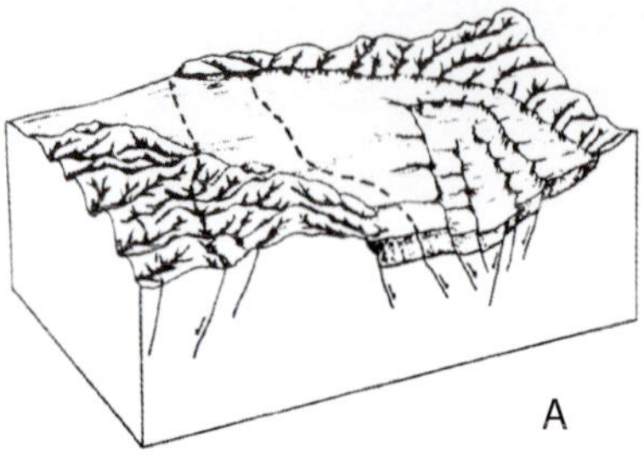

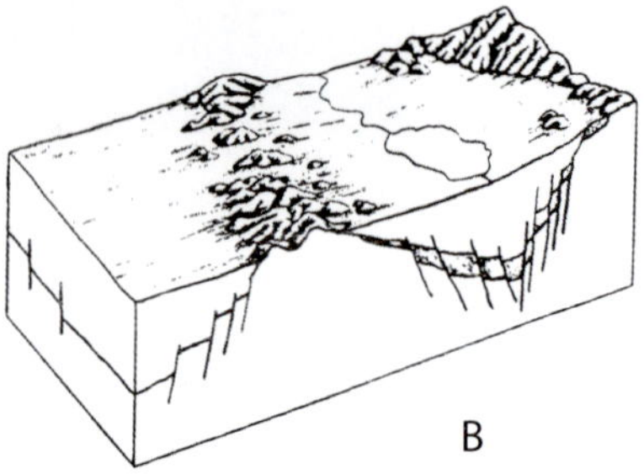

Two stages in the evolution of a Basin and Range landscape. Block A (Miocene, early Pliocene) shows an early stage when extensive normal faulting has created a complex down-dropped fault block (graben) flanked by two relatively up-thrown blocks (horsts). The floor of the basin contains early deposits of basin-fill that are also faulted by continuing subsidence. Block B (late Pliocene, early Pleistocene) shows a later stage after faulting has ceased. Erosion has reduced the heights and extent of the ranges and the basins are completely filled by sediments. The existence of a playa lake indicates that the basin has internal drainage. Small isolated hills called inselbergs are the eroded remnants of earlier uplands. Modified from Menges, 1981 (AGI) and Menges and Pearthree, 1989.

Basin features of the Altar Valley. This representational view shows the normal faults along which the crustal blocks of the Altar Valley have subsided relative to the Baboquivari Mountains block. The dashed line suggests the approximate trace of the border fault. Looking at the front of the block, the lowest layer is faulted bedrock and the middle layer represents early basin-fill (late Miocene to early Pleistocene). The upper layer represents middle Pleistocene to Holocene fluvial sediments upon which the modern landscape features have been created. Terrain image © Google Earth 2012.

After faulting and subsidence diminished, the topographic relief gradually reduced as the mountains continued to erode and the basins received sediment. At the same time the drainage evolved from a closed basin system, like that of central Nevada, to the integrated system of today that drains from the Santa Cruz and San Pedro watersheds into the Gila River and then into the Colorado River.

Each valley has evolved in a unique way, depending on many geologic, geomorphic, and climatic factors. The landforms of the Santa Cruz River Valley with its relatively narrow width, relicted alluvial fans and stream terraces, and adjacent steep mountain fronts show a complex history of fluctuations in fluvial (river) processes. The valley is an excellent example of a Basin and Range river system that has evolved in concert with major climatic and tectonic fluctuations.

The Santa Cruz River Valley between Tucson and Nogales reveals many clues to the river's history. Prominent are alluvial fans of sediment that was washed from canyons to form an apron of fans called a piedmont or bajada. The piedmont has been extensively dissected by streams as climatic and fluvial factors changed. The Madera fan (see photo below) is a classic alluvial fan that has developed in an arid/semi-arid Basin and Range setting.

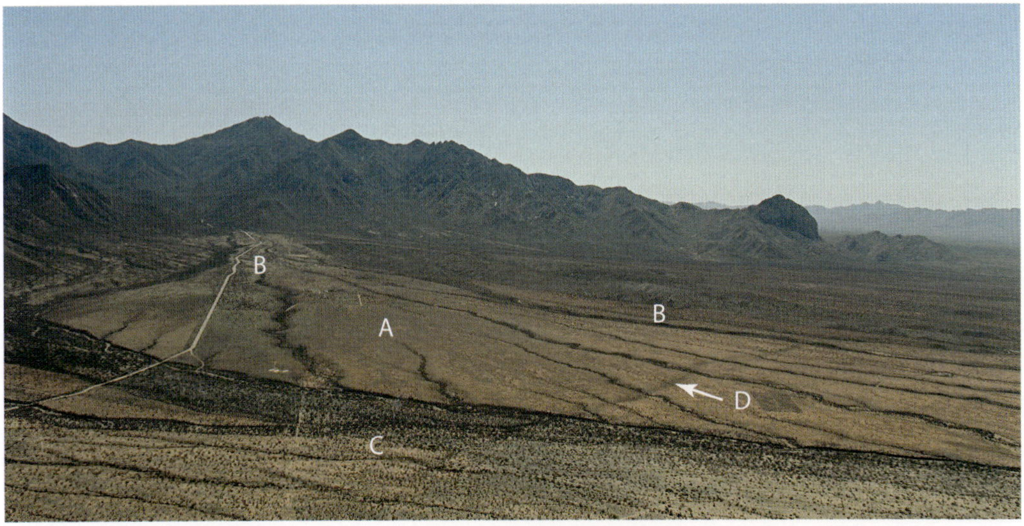

The Madera alluvial fan. This classic example of a large dissected fan occurs at the mouth of Madera Canyon on the west flank of the Santa Rita Mountains near Green Valley, Arizona. The present drainage is near the distant edge of the fan below the ridge and Elephant Head Rock. An abandoned section of the fan (A) is dissected by several channels (B) that radiate out from the mouth of Madera Canyon. The Madera fan is truncated by the development of an adjacent younger fan (C) seen in the lower third of the image. A faint dark line on the fan surface (D) is the trace of the fault that ruptured most recently. Image courtesy Mark Heitlinger, University of Arizona – Santa Rita Experimental Range.

The Altar Valley has a different story. Today there is about 4000 feet of relief between the Baboquivari range crest and the floor of the Altar Valley. Beneath the valley floor as much as 4800 feet of sediment has been washed from the surrounding ranges (Richard, Stephen M., et al., 2007, *Estimated Depth to Bedrock in Arizona*). That means that there may have been about 9,000 feet (2740 m) of displacement between the Altar Valley and the Baboquivari Mountains. The valley is about 50 miles long with its upper reaches just north of the Mexican border at Sasabe. North of Three Points (Robles Junction) it becomes the broad and nearly featureless Avra Valley.

The landscapes in this area are in contrast to the Santa Cruz River Valley. The wealth of diagnostic landforms is minimal, because the Altar Valley is much less evolved. There is no large axial stream like the Santa Cruz River, but the valley is drained by Brawley Wash which flows north and continues through the Avra Valley to the Santa Cruz River.

Locally the name Brawley Wash refers to the part of the drainage that has a deeply incised channel that is up to 20 feet (6 m) deep and 1500 feet (457 m) wide. In the late 19th century the axial valley floor was undissected and water from flood events was able to spread across the floodplain with minimal erosion and infiltrate the soils. Drought, overgrazing, and other human activities created a situation that allowed water from a breach of a ranch reservoir in 1904 to incise a channel. It has enlarged and become a serious land use and ecological threat to the valley. Today, Brawley Wash is the focus of major erosion control and restoration efforts by local and governmental entities.

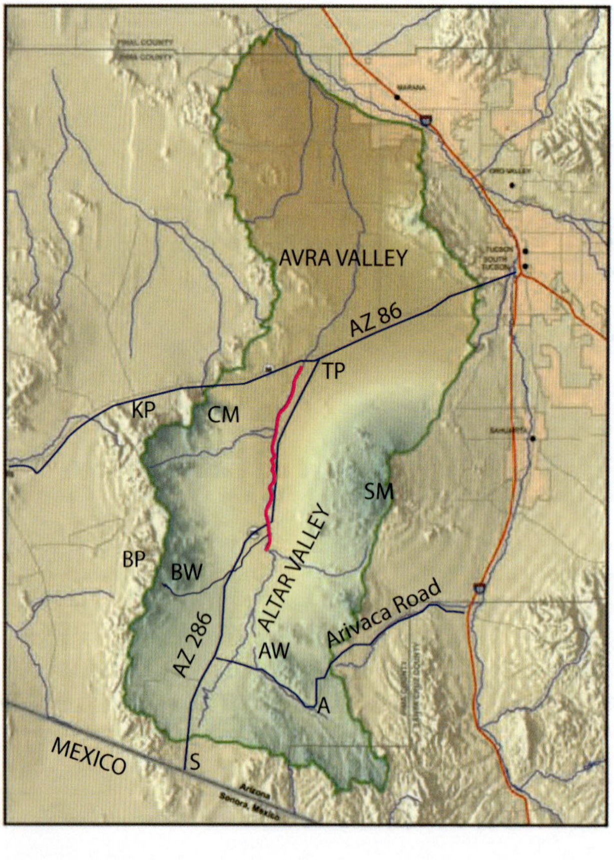

Left: the Brawley Wash watershed. The drainage originates a few miles north of the border with Mexico and flows north. Upper tributaries combine to form an established axial stream in the reach below Arivaca Wash (AW). Brawley Wash (BW) leaves the Altar Valley and enters Avra Valley near Three Points (TP). The reach of Brawley Wash most seriously impacted by recent arroyo cutting is shown in pink.

Other labeled features are Baboquivari Peak (BP), Brown Wash (BW), Coyote Mountains (CM), Kitt Peak (KP), and Sasabe (S). Map from NRCS.

The diagrams below depict three stages in the development of the landform features seen at the mouth of Brown Canyon. The term "grade" refers to the level below which a stream cannot cut. Degrade means the stream is cutting downward and aggrade means it is depositing sediment and filling its valley.

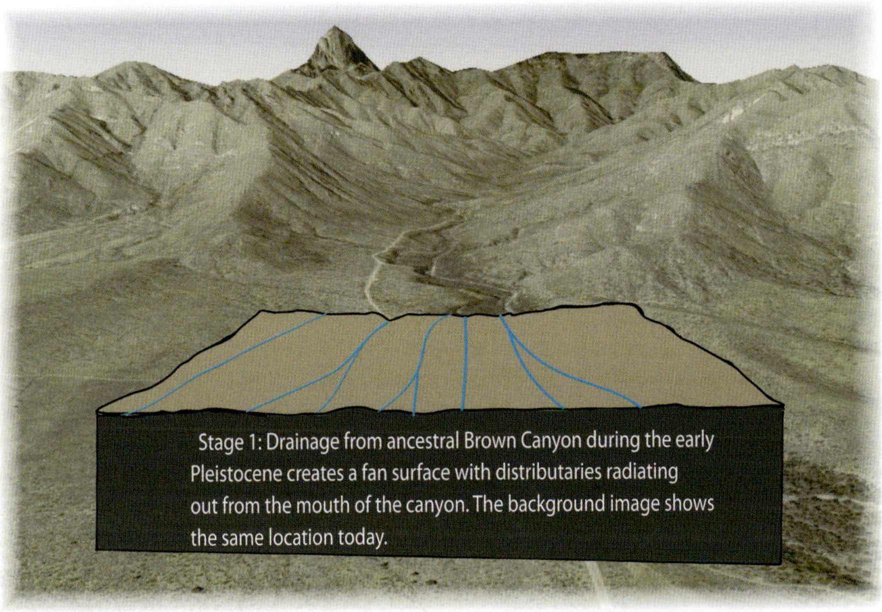

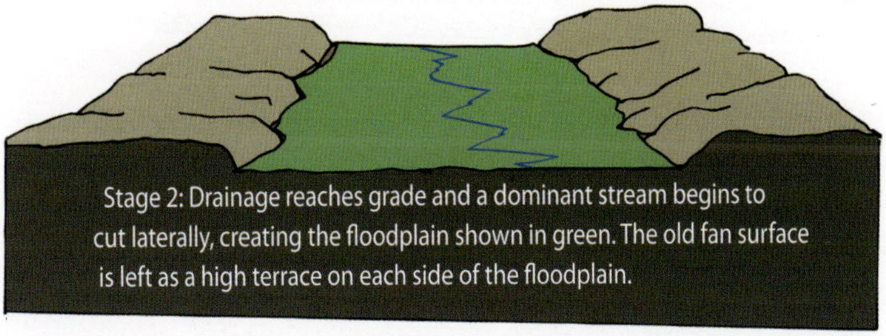

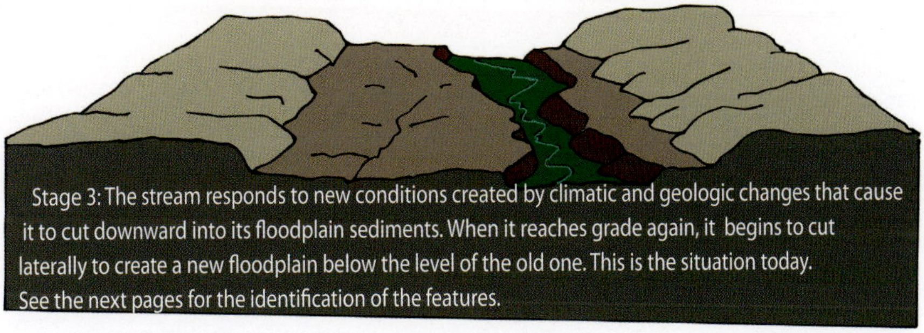

The image to the right shows the geomorphic setting of Brown Canyon (see pages 7 and 8 for a wider view). The canyon is deeply incised into the east slope of the Baboquivari Mountains and has built a major alluvial fan into the Altar Valley. Adjacent fans include the Thomas Canyon fan and the Chiltepines fan which contribute to the Baboquivari Mountains piedmont in this area. Note the variation in the color of the soils and the different drainage patterns on the different fans.

The image below shows a detailed view of the mouth of Brown Canyon. Jurassic plutonic rocks form the steep hillslopes above the piedmont. There is no evidence of a basin border fault. The slope break at (A) corresponds to the extent of erosion of the bedrock as the mountain front retreats. From the mountain front an early to middle Pleistocene fan extends basinward. Remnants of the surface (C) are commonly covered with reddish soil, distinctive of older surfaces. Late Pleistocene channels (B) were incised into the fan during an episode of degradation. During this time a pair of fan terraces was created. They are visible on both sides of the mouth of Brown Canyon. This guide's Rock-Stop 2 is located on an isolated remnant of the surface. With greater weathering of the fan, some areas between channels take on a rounded whale-back shape (D). These ridges are appropriately called ballenas (Spanish for whale) and are distinctive of old and extensively eroded fans. The present stream shows the meandering shape of a graded stream and occupies a floodplain (E) it has created by lateral cutting. The series of concave cutbanks (F) are evidence of the time when the stream was at grade and widening its floodplain. Later a small younger terrace (G) was created and bounds the present floodplain. The modern stream appears to be at grade with a balance of aggradational and degradational processes. In the present arid/semiarid climate virtually all geologic work is done during flash flood events when large amounts of material are moved. This is especially evident in the upper reaches of the streams.

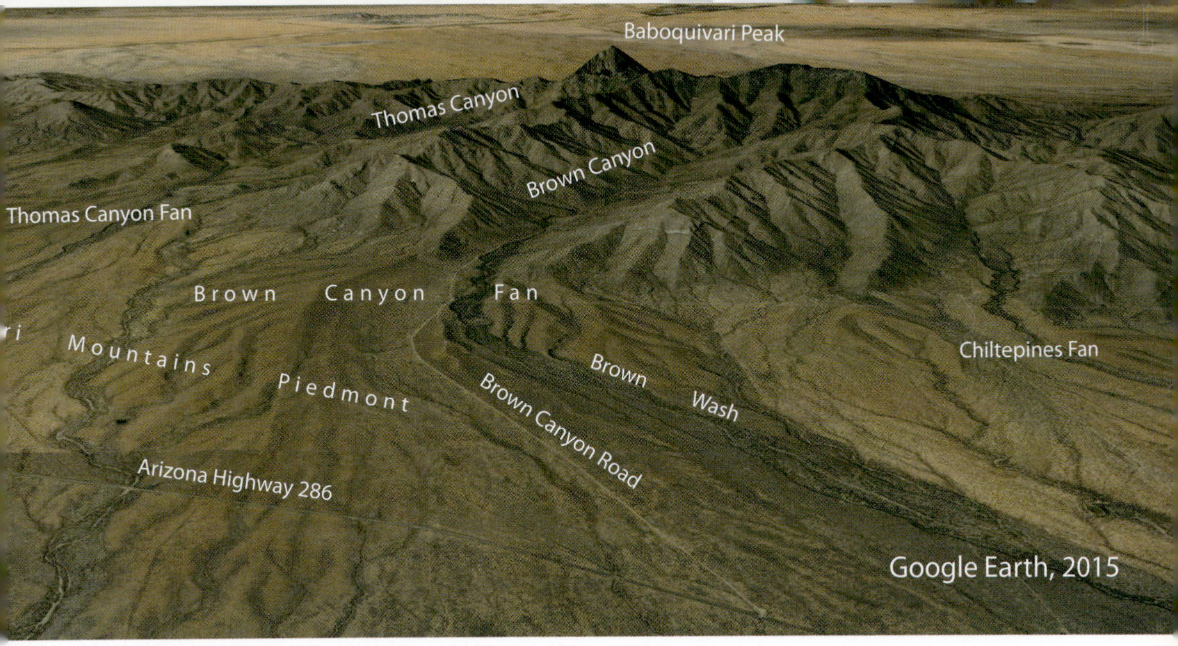

Below is an aerial view looking east from above the mouth of Brown Canyon. Brown Wash and many smaller washes drain eastward across the fan surface and are tributary to Brawley Wash, located approximately where the light color of the fan ends. A ranch dam crosses Brown Wash near the bottom of the photo and Rock-Stop 2 can be seen above and southeast of it. The mountains on the left horizon are the Sierrita Mountains and the Santa Rita Mountains are on the right.

The causes of fluctuation in stream behavior are difficult to determine. The most common explanation calls upon climate change. Another important potential cause involves changes in base level, which is the lowest level a stream can cut down to. For the Colorado River it is sea level, for a stream flowing into a lake, it is the lake level. For a stream flowing into a closed basin, as in the Basin and Range, it is the floor of the basin.

During the erosion of a region, divides between watersheds become lower and two things can happen. A basin can fill with sediment, water can spill over the divide, and then drain the upper basin into the lower one through the new channel. The base level for the upper basin now becomes the lower basin floor. The other process is called stream capture. A stream draining into a lower basin can gradually erode headward, breach the divide between the two, and capture the upper basin drainage.

Although Basin and Range faulting ended about five million years ago, all tectonic activity has not ceased. The Madera fan sediments were faulted within the past few thousand years (see photo on page 44) and there was a large earthquake in 1887 in northern Sonora, Mexico, southeast of Douglas, Arizona. Uplift or subsidence of the crust can raise or lower the streams relative to base levels and the streams would adjust to the new control.

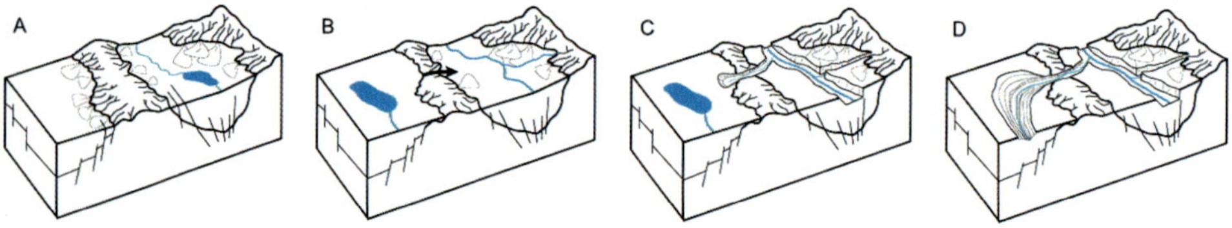

Basin integration by spillover and stream capture. (A) Two adjacent basins after the end of subsidence. Both have internal drainage. (B) Filling of left basin with sediment causes spillover into right basin. (C) Right basin stream captures the drainage of the upper basin and lowers its base level. During the same time the right basin has become integrated with a downstream basin which lowers its base level and results in downcutting of its stream to the new base level with formation of floodplain and stream terraces. (D) Upper basin stream cuts into its basin-fill as it incises to a new base level. This is a possible scenario for the integration of the sub-basins of the Gila River system into the modern drainage. Image courtesy Matthew Jungers, from Arizona Geology Magazine, December 2012, modified from Menges and Pearthree, 1989.

Each of these processes helped to integrate separate basins into a network of interconnected basins. Each basin experienced changes in discharge and base level that could change an aggrading (depositing sediment) stream to a degrading (eroding) one or just the opposite.

The climate in the Southwest during late Miocene and Pliocene times was warm and more humid than today and stream systems were capable of significant erosion and transport. This was the time when Basin and Range extension and subsidence ended and the mountains were being vigorously eroded. Much of the sediment that occupies the basins was deposited during this time. The alluvial fans that create the piedmonts were deposited at the end of the Pliocene and early Pleistocene.

The climatic change from a stable Pliocene climate to the fluctuations of the Pleistocene

Sopori Wash between Arivaca and Amado exhibits arroyo cutting that is typical of many streams in southeastern Arizona. Stream dissection similar to this began about 8,000 years ago and continues today.

radically altered landscape evolution in southern Arizona. There were no glaciers, but during glacial episodes it was much moister, stream systems were more active, and the presence of vegetated slopes slowed erosion. During interglacial times, the climate became semiarid and even arid, resembling today's conditions. Transitions between relatively wet and dry conditions such as occurred during the end of the Pleistocene may have resulted in substantial hillslope erosion and deposition of sediment in the valleys. Streams began to cut downward, creating the incised drainage seen on the older fan surfaces. Climate fluctuations caused the streams to vary from degradational to graded to aggradational many times.

At the same time, the drainage system was evolving and becoming more integrated as formerly closed valleys became part of the larger system. Base levels were changing, discharge was changing, and new geologic conditions appeared, all affecting the entire system.

Today is an interglacial time with active climatic and fluvial changes. In the latest stage of landform development, an episode of arroyo cutting in southeastern Arizona began about 8,000 years ago followed by many more episodes. Human activity since the late 19th century has abetted the process, as we have seen at Brawley Wash.

A tributary of Brawley Wash has dissected a deep channel that can be seen where it crosses AZ 286 eight miles (13 km) north of the Brown Canyon road. The erosion seen at this location has taken place since the early 20th century.

A CHRONOLOGY OF EVENTS IN THE EVOLUTION OF LANDSCAPES NEAR BROWN CANYON	
Miocene - middle Pliocene	Subsidence of the Altar Valley Basin along faults that are now buried under basin-fill. Contemporaneous erosion of the adjacent highland areas and deposition of early basin-fill. End of faulting and subsidence about 5 Ma.
Late Pliocene - early Pleistocene	Establishment of early basin drainage that was probably initially internal. Continued filling of basin with later basin-fill sediments and formation of the early alluvial fans. Climate transitioned from more stable and humid Pliocene conditions to highly variable Pleistocene climates.
Middle Pleistocene - Holocene	Continued erosion of uplands and formation of the piedmont by slope retreat and formation of later fans. Variable stream behavior lead to dissection of the older fans and creation of tributary floodplains. The floodplain and stream terrace at the mouth of Brown Canyon were formed during this time. Drainage integration with the Santa Cruz River led to establishment of the Brawley Wash drainage system.
Recent	The current interglacial time with onset of arroyo cutting about 8,000 years ago. Settlement by Europeans in 19th century with extensive modification of landscapes, ecosystems, and stream systems. Major erosion of Brawley Wash after 1904.

GEOLOGIC AND TRAILS

LOG FOR THE ROADS OF BROWN CANYON

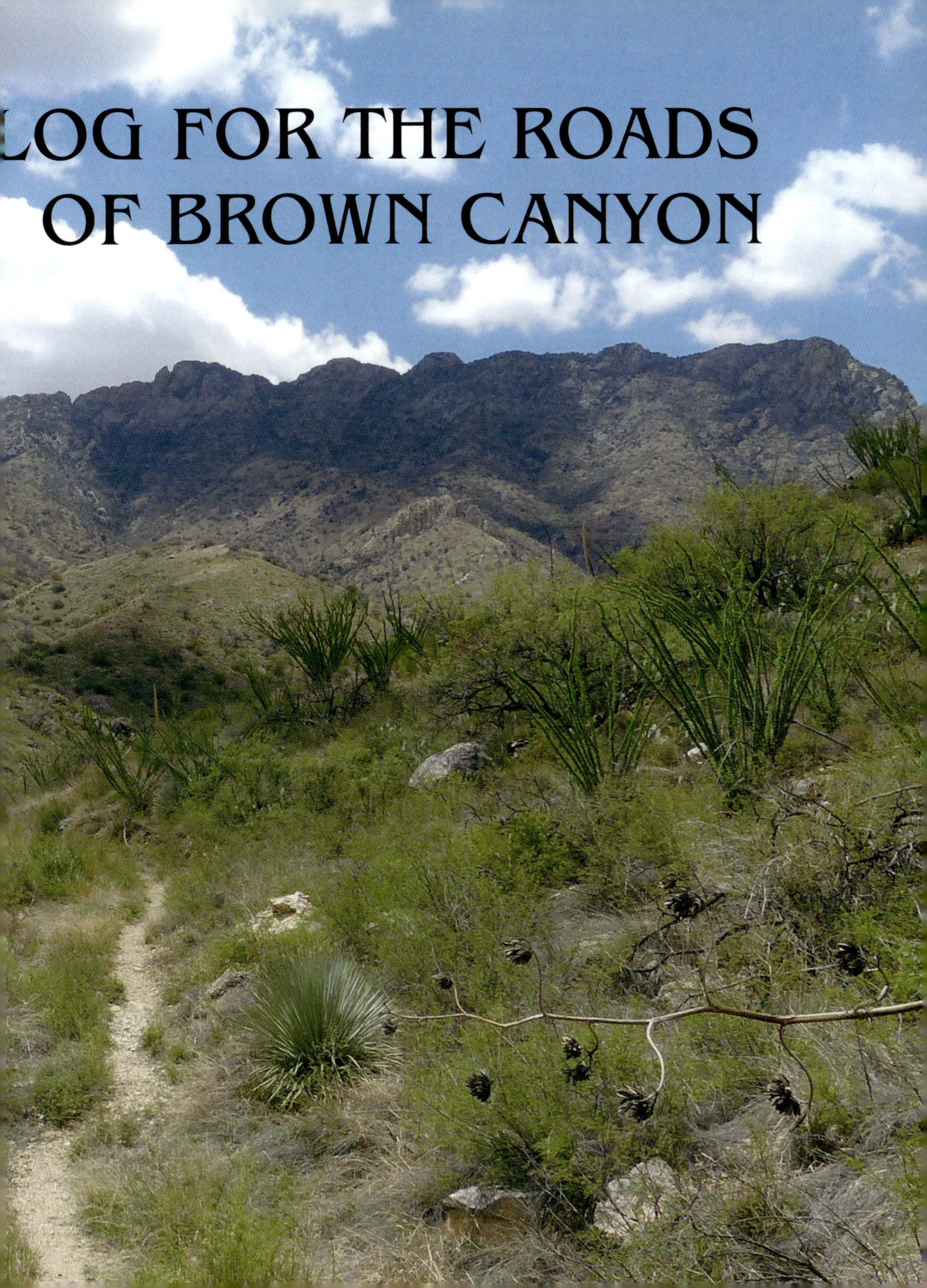

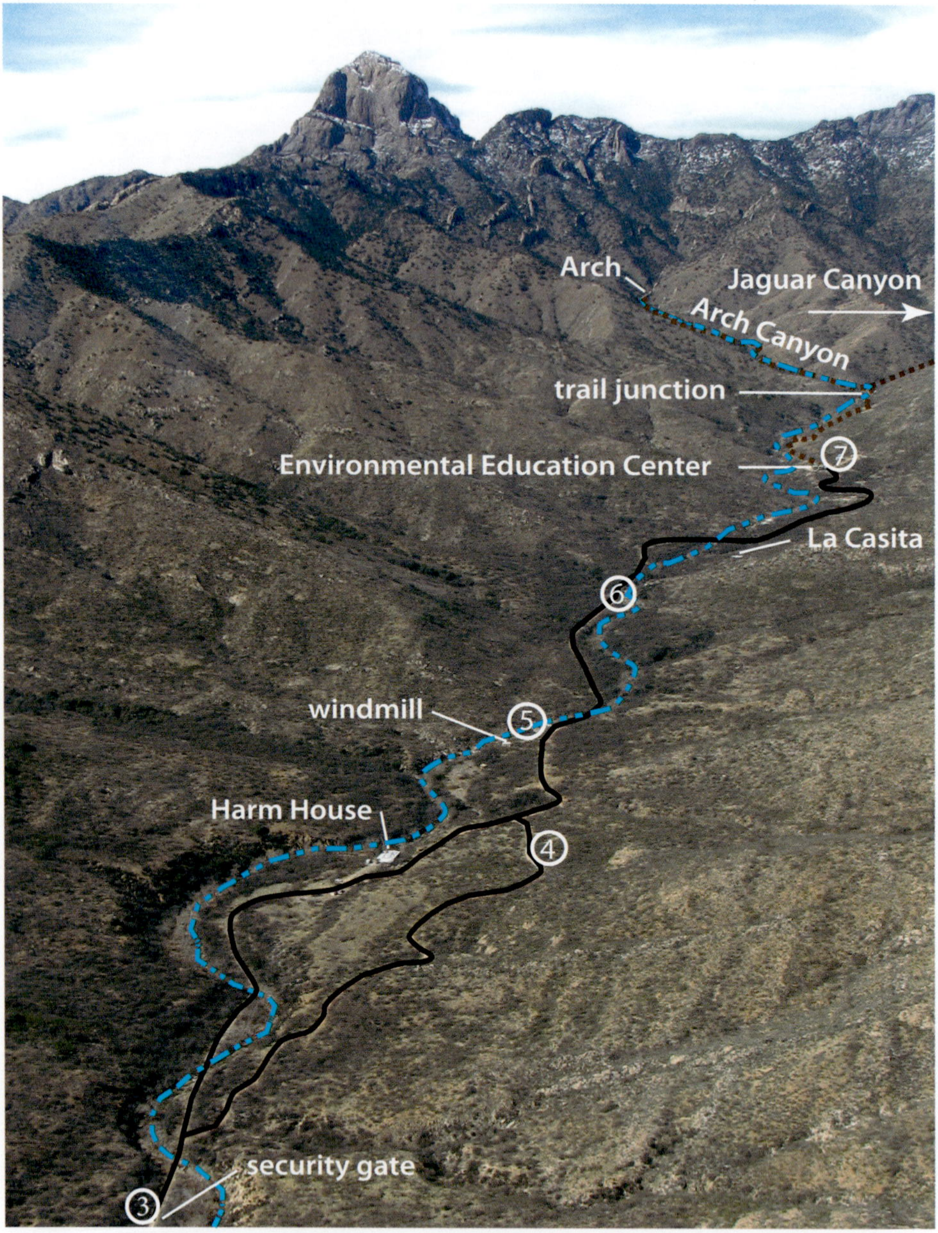

Brown Canyon and Baboquivari Peak

Previous pages: view of head of Brown Canyon from trail above the Environmental Education Center. This photo also appears on page 65 where the features are labeled and discussed.

The Baboquivari Mountains are, with few exceptions, wilderness. Access is limited and can be challenging. Of course, this is one thing that makes these mountains so special. The few roads are mainly jeep and ranch roads. Several of the ranch roads that have public access begin as graded dirt and become four-wheel drive roads as the mountains are approached. A well graded dirt road leads to Brown Canyon. To travel the road within the canyon, high clearance and good tires are advised.

The Brown Canyon unit of the Buenos Aires National Wildlife Refuge has a maintained trail system. All other trails are ephemeral, having been established by migrants, ranchers, miners, hunters, and wildlife. This log is for roads and maintained trails.

Each log section for a road or trail begins at a location which has a physical description and GPS coordinates as UTM with a WGS84 datum. Log distances are given in miles and (Km) from the log start to a particular rock-stop. The entry for a rock-stop contains a physical description of the place, GPS coordinates, descriptions of geological features and a discussion of what they mean. Photos and diagrams will help you locate features and assist in understanding their meaning.

Each log section and rock-stop is located on the geologic map. Refer to the map inside the back cover and to the cross-section on pages 23-24.

An aerial view of Brown Canyon and Baboquivari Peak. Features used in the log and shown on the geologic map are labeled.

The rocks in the lower third of the photo are Pavo Kug Granite (Jp). The Tinaja Springs Porphyry (Jt) is exposed in the center third of the photo. In the canyon bottom the contact is between Rock-Stops 4 and 5. Allison Camp Rhyolite dikes can be seen as craggy light-colored bands trending across the hillsides. The slopes above the trail junction expose the Chiltepines Member of the Pitoikam Formation (Jc). The Mulberry Wash Formation (Jm) occupies the middle third of the canyon headwall, approximately from the word "Jaguar" up to the snow. The snow-covered rocks are part of the Baboquivari Peak pluton.

The geologic map shows the approximate location of contacts as established by reconnaissance mapping, but it is difficult to place them on the photograph with any certainty.

ROCK-STOP 1: junction of AZ 286 and Brown Canyon Road

mile marker 20.5, log mile 0.0, 12R E 0455191 N 3514438

To get to this place you drive AZ 286 through the Altar Valley, a broad valley that extends from Three Points Junction deep into Sonora, Mexico. Here it is flanked on the west by the Baboquivari Mountains and on the east by various smaller isolated ranges.

This landscape of mountains and broad valleys is a consequence of crustal stretching and block faulting. Each range is a fault block that remained higher than the adjacent valley block during faulting. The basins are sometimes thousands of feet lower than the ranges and filled with sediment washed from the ranges. See the Sky Island map on page 5 for the geography of the area.

Modern southern Arizona landscapes have emerged from the unique event geologists call the Basin and Range extension. This remarkable tectonic episode stretched and faulted the western North American crust to the breaking point and created a mosaic of broken fault blocks. This all began to evolve about 30 million years ago, and faulting lessened about 5 million years ago. Through several million years and countless earthquake events, basins subsided as the crustal rocks dropped along faults that bordered the basins. In a single slippage the rocks may have moved several feet. After innumerable drops the total movement could add up to thousands of feet. The Altar Valley basin floor down-dropped as much as 10,000 feet (3000 m). As extension lessened about 5 million years ago, earthquake events became less frequent. The last significant earthquake of this type occurred in 1887 in northern Sonora, Mexico, south of Douglas, Arizona. Geologists think that such earthquakes in this region are now happening every few thousand years.

Look at the scene before you. The Baboquivari Mountains are very rugged and Brown Canyon is an impressive canyon descending from the steep and wild headwall near Baboquivari Peak. The drainage from the canyon then flows away from the mountain front, across a gently sloping surface to the basin floor of the Altar Valley (see pages 47 and 48). From here, washes drain northward to the Gila River and then to the Colorado River. For hundreds of thousands of years this drainage has been very intimately integrated and ultimately controlled by the entire stream system all the way to the Sea of Cortez.

The Baboquivari Mountains from the east side of the Altar Valley. Mildred Peak is far left, Kitt Peak far right, with Baboquivari Peak in the center. Brown Canyon is below the horizontal section of the divide at center.

Watch for these mailboxes at Brown Canyon Road.

Early erosion of mountains and basin filling began as soon as there was sufficient topographic relief. Later, the climatic changes of the Pleistocene Epoch accelerated weathering and erosion. There were no glaciers here, but the climate was much moister during glacial epochs and drier during interglacials, resulting in more active stream erosion and deposition. As flooding streams left the confines of the mountains they fanned out across the basin carrying and depositing sediment. The resulting landscape feature is called an alluvial fan. Sloping gently away from the mountain front is an apron of overlapping fans called a bajada
or piedmont. This is a common feature of arid/semi-arid Basin and Range landscapes. The Sierrita Mountains piedmont, to the northeast, is a classic example (see map on page 5).

As you drive up to the next stop, you will be driving across alluvial material that has been eroded from Brown Canyon at various times during and after the Pleistocene Epoch.

ROCK-STOP 2: emergency beacon, mile 2.6 (4.2 Km), 12R E 0452252 N 3513059

Park at the foot of the track road to the emergency beacon and walk up to the top of the hill. From the highway, you have driven up the piedmont across relatively young alluvial deposits. The surface you are standing on now is underlain by a remnant of a Pleistocene alluvial fan (pages 47 and 48). These fans can best be seen in an aerial view (pages 7 and 8). Look across the wash to the north and to the south and you can see the same high, gently sloping surface of alluvial deposits. But notice that the modern Brown Canyon stream is well below the level of the fan surface. The fans were made during wetter Pleistocene glacial intervals. The modern climate is an interglacial one that is more arid. More active erosion has caused the streams to erode valleys into the fans. In a climate like today the geological work of running water occurs during flood events. Within the last few thousand years arroyo cutting began in response to climate changes. Since then human activity has exacerbated the erosion. Brawley Wash and its drainage are now being actively managed to reduce erosion in the Altar Valley.

The mouth of Brown Canyon from the emergency beacon. Red lines outline several prominent Allison Camp Rhyolite dikes. The white line marks the same fan surface you are standing on.

From here you can start to see some geologic details of the mountain front. One of the most distinctive features around Brown Canyon is the array of igneous dikes. Review the section on igneous rocks, especially the block diagram on page 15. Compare the photo above with the mountainside to help you recognize the dikes as seen from here. Find the dikes on the guide map and watch for more dikes as you continue.

Return to your vehicle and continue the drive into Brown Canyon. As the walls close in watch for the dikes—craggy light-colored walls and ledges on the hillsides. Find yourself on the geologic map and you will note that lower Brown Canyon is underlain by the Pavo Kug Granite (Jp) with cross-cutting Allison Camp Rhyolite (Ta) dikes. The valley bottom and lower slopes are covered with rocks that have been transported down the hillslopes by gravity and water. Bedrock is largely buried by this material, but there are several outcrops close to the road that you can walk to and take a close look at the bedrock.

ROCK-STOP 3: security gate, mile 3.8 (6.1 Km), 12R E 0449437 N 3510110

Here you are in the narrow canyon bottom of an active stream. The canyon sides are steep because of the resistance of the rocks to weathering and erosion. In a climate like this most of the geological work of running water occurs during flood events. Flash floods can be very powerful and can move

The security gate. The fan surface is at the top of the bluff above the gate. Exposures below the surface are stream and debris flow deposits.

many thousands of tons of material in minutes. In steep terrain like Brown Canyon, the fast-moving water with its sediment load can transform into a debris flow that is often a fast-moving fluid mix of water and sediment that can transport huge boulders long distances. We can see examples at this stop.

Once through the gate, park your vehicle and make your way to the right (north) through the brush to the wash channel. The sediment in the active channel contains rocks from the entire watershed above here, so you may get a sample of all the rock units of Brown Canyon. The modern stream has dissected old stream and fan deposits as old as the Pleistocene. Look at the boulders in the cliffs and imagine the power required to move that material! Compare it with the modern deposits you are standing on. The surface at the top of the cliff of boulders corresponds to the same fan surface we saw earlier.

Return to your car. After you leave the gate area, you will cross a wash on a concrete pad, then take a right fork up a steep hill and find yourself on a surface providing good views of the lower canyon.

Debris flow deposits near the security gate. These sediments are the upper part of the fan that is at Rock-Stop 2 and extends into the valley. The fan surface is at the top of this exposure.

ROCK-STOP 4: upper Brown Canyon Road, mile 4.1 (6.6 Km), 12R E 0448993 N 3515318

You need to find the right spot here, so study the photo and the Google Earth image below. Now look around you at the hillslopes, crags, and the high cliffs—all are igneous rocks. Look up-canyon to its head and down to where you entered—nearly all the canyon you can see from here is igneous. The only exceptions are the rocks at the lower-most part of the slope at the back of the canyon which are sedimentary.

These rocks formed about 170 million years ago beneath the margin of the growing continent. If you had been there you would have found yourself in a fiery foundry that was manufacturing the stuff of the North American continent. The spiny prickly vegetation on the slopes above us and across the valley largely conceals a complex mix of stock-like masses and interlaced dikes.

Exposed in the roadway is the most extensive of these rock masses, the Pavo Kug Granite (Jp). The photos below show what to look for as you walk the road. Poke around in the loose rocks above the road and you will find other examples of the Pavo Kug. Use the illustrations in the first part of this guide to help you identify the minerals and understand how the rock was formed.

The house below belonged to Ray Harm, a well-known wildlife artist who was instrumental in making it possible for the Refuge to acquire this property.

Continue driving past the fork to the Harm House. Watch for a windmill on the left.

Looking upslope to the north from Rock-Stop 4. Pavo Kug Granite (Jp) in foreground.

Pavo Kug Granite exposed on the road at Rock-Stop 4. White feldspar and grey quartz are easily identified. This granite is unlike typical granites because it is leucocratic (low in content of dark minerals).

ROCK-STOP 5: windmill and lower tinaja, mile 4.5 (7.2Km), 12R E 0448643 N 3514536

The complexity of the intrusive rocks is best appreciated on the steep hillslopes and ridgetops where only the heartiest of hikers venture. But just upstream from the Harm House the stream has cut through a striking display that is easily accessible. Drive from Rock-Stop 4 past the fork in the road that leads back down to the Harm House, and then past a windmill and tan water tank. A glimpse down into the wash will reveal the rocks featured at this stop. Drive across the next wash and find the nearest spot where you can park off the road. On the north side of the road is a very light-colored somewhat glittery rock. This rock is the common dike rock called rhyolite.

Make your way down to the wash and the big flat slab in the middle of the stream bed. On your way look at the diversity of rocks in the stream bed. Especially look at the perfectly rectangular crystals in many of the rocks. These are feldspar crystals and the rock is called porphyry because these crystals are distinctly larger than the crystals in the surrounding matrix.

The flat rock (photo below) is the bedrock and is part of a dike system. This is only one of several different types of dikes found in the canyon.

The most obvious features you see are the black blobs of all sizes that are imbedded in the rock. The blobs suggest the mingling of the felsic and mafic magmas, similar to the blobs in a lava lamp.

Left: outcrop of felsic dike with comagmatic blobs.

Lower left: intergrown feldspar crystal in a blob.

Below: feldspar crystals that have been replaced by epidote during later alteration of dike rock.

The course of the wash is through a cascading series of rock pools lined with cattails and other aquatic plants.

If two magmas have very different properties or chemical compositions, they may be prevented from mixing and forming a single magma. So imagine blobs of mafic magma floating in the felsic magma along with the solid feldspar crystals that are floating and growing in both. We can call them "comagmatic blobs".

The next outcrop downstream is more of the dike we looked at on the road. Chemical weathering of these light-colored, nearly white rocks makes them appear to be rusty or even black. As you continue downstream the white rock occurs until you reach a clump of trees and brush in mid-wash. Make your way down the left channel. An old ranch dam crosses the wash here. Continue walking to the edge of a cliff and you can see the dam. Below the dam the stream flows through a beautiful series of rock tanks called tinajas. During the wet season

Three kinds of dike rocks. The white rock occurs on either side of this display as well as nearby. The greenish rock contains millimeter-sized grains of a mafic mineral. The grey rock on the left is uniform in appearance. The contacts between the rocks are parallel and there are no obvious relative age relationships between the dikes. A careful examination of the contact zones suggests that the mafic rocks are younger.

they are filled with water, cattails, and wildflowers. Water remains here through all but the driest years. Butterflies love this place.

The lower photo on the previous page is looking upstream. The familiar white rock is here with two parallel bands of a greenish rock that contains small pervasive dark splotches. Another rock at the far left is a uniform grey. Which rocks you will actually be able to see here will depend on how the last flood covered the channel with boulders and gravel.

There are three different rocks here. Because they are fine-grained, they are either volcanic or subvolcanic rocks. The contacts are intrusive, so we can conclude that they are dike rocks. But which intruded which? The ideal way to know would be by using the idea of cross-cutting relationships introduced on page 16, but here all the contacts are parallel and so, more subtle evidence must be used. It requires an expert and experienced eye to see the clues. The verdict is that the dark-colored rocks have intruded the light-colored rocks.

ROCK-STOP 6: La Casita dike, mile 4.8 (7.7 Km), 12R E 0448127 N 3514727

Park near the minor wash that crosses the road at mile 4.8 (watch for a pair of refuge boundary signs) and make your way down into the Brown Canyon stream channel. Walk up the channel to where bedrock blocks the way, noting the gorgeous rocks lying in the stream channel.

We've seen that the kinds of bedrock upstream are represented in the rounded water-worned rocks in the channel. This spot contains a delightful assortment of these rocks. The photos below and on the next page identify several.

A major rhyolite dike crosses the canyon at this place. The massive outcrop on the north side is water-worn and the mineral grains are readily seen. Look for feldspar and quartz. The dike is easy to trace as it ascends the slope to the north.

The massive rock on the south side is the same dike, but looks much darker because of chemical weathering.

Return to your vehicle and continue up to the Environmental Education Center and the end of the road.

Water-carved notch in dike. The large boulder in the notch is derived from the Mulberry Wash Formation.

Cobble of Baboquivari Peak Perthite Granite. Pink crystals are perthite feldspar.

Cobble of Mulberry Wash Formation. The angular to subangular clasts are rhyolite.

ROCK-STOP 7: Brown Canyon Environmental Education Center,
mile 5.5 (8.9 Km), 12R E 0448993 N 3515318

This splendid structure started out as the stone cookhouse for a ranch and grew through several owners to a family mountain retreat. It is built on Tinaja Springs Porphyry (Jt). This is our next major rock unit and it is extensive in the middle part of the canyon. Look at the map and cross-section. Park on the uphill side of the building and walk to the trailhead next to the propane tank and generator shed. The trail is a former ranch road that leads downhill between rock walls. The walls were built by ranch laborers in the 1950s and they contain all the rock types of the upper canyon.

Walk down the road a few meters and examine the rock that forms the surface of the road. The Tinaja Springs Porphyry is typically a rust-colored rock with tiny glittery crystals of quartz and flashing cleavage faces of feldspar. The rock has a porphyritic texture with larger crystals of feldspar. Its form is a large sill-like injection of rhyolite and it is about 170 million years old.

Looking down the trail from theEnvironmental Education Center. The Tinaja Springs Porphyry is exposed on the surface of the roadway. Note the distinctive rusty appearance.

Detail of Tinaja Springs Porphyry. White crystals are feldspar and grey crystals are quartz.

Before you leave the area, check out the Brown Canyon rock collection at the picnic tables behind the building. The next section of the log starts here as a rock-walk.

ROCK-WALK 1: Education Center to Arch Trail - Jaguar Trail junction,

0.6 miles (1 Km)

Stroll between the walls, seeing how many kinds of rocks you can recognize. A hundred meters beyond the main stream crossing, a white boulder on the left side of the trail will reveal cavities filled with quartz crystals. The Allison Camp Rhyolite dikes are very silica-rich. Cavities in the cooling dike were filled with a silica-rich solution and quartz crystallized, lining the cavity walls.

The abandoned Jantzen house belonged to the last family to ranch the canyon. From here are two route options. One follows a trail alongside the stream and the other goes up along the old ranch road. This log will take the high road to see and discuss the spectacular vistas and then return down the stream. For a pleasant detour, you can take the low route from here by going around the west end of house and finding a trail that crosses the stream.

Continuing on the ranch road, walk uphill alongside more rock walls. As you continue along the trail the view of the canyon's headwall unfolds. Walk to a spot where you can see the view in the photo below. The major features are identified. Look at the section on magmatic features (page 15) to understand what you are seeing.

Baboquivari Peak is part of a larger stock. It stands high because post-intrusion chemical alteration added silica to the rock and increased its resistance to weathering and erosion. The rhyolite dikes are also more resistant than the surrounding rock and stand high as walls, buttresses, and towers. The trail is on Tinaja Springs Porphyry, but most of the outcrops you see are dike rocks.

Continue walking past a stone shed to the site of the earliest remaining ranch house and facilities in the canyon, the Perkins Ranch. On the way, examine the impressive boulder of Mulberry Wash Formation lying in the wash channel (photo top of next page).

The Jaguar Canyon Trail goes around the cattle loading chute and water tank. The Arch

Brown Canyon's west wall. We can see three episodes of magmatic activity here. The oldest is the 170 Ma Mulberry Wash Volcanics. The next is the intrusion of the 146 Ma Baboquivari Peak pluton. The youngest is the 26 Ma Allison Camp Rhyolite dikes. A larger version of this photo is on pages 51 and 52.

A boulder of Mulberry Wash Formation. Note the evidence for bedding in the masses of lighter colored clasts that are inclined to the left.

Trail continues up the main canyon. This log returns back down the trail, past the stone shed, to a lesser used trail that descends the slope to a lovely set of tinajas (12R E 0446702, N 3515725). You will be fortunate if you visit this place when the stream is running and the tinajas are connected by waterfalls. The rock outcrops here are great for eating lunch and discussing rocks. The rock here is fine-grained and reveals virtually no clues to its origin, but is part of a dike that crosses the canyon.

A few meters up the stream to the right of the next tinaja, a sharp eye will spot the only non-igneous rocks in the canyon—the siltstone, sandstone, and conglomerate of the Chiltepines Member of the Pitoikam Formation.

Geology students resting on a felsic dike rock at the tinaja below the Arch Trail/Jaguar Trail junction.

The circle shows the location of a group of rounded pebbles in an otherwise uniform siltstone. The location is the next pool above where the group photo was taken.

Find the trail from the tinaja that continues down the stream on its south side. Walk past an abandoned utility trailer to a wash crossing (12R E 0446807, N 3515625). Walk up the wash a few yards and you will find fine-grained sedimentary rocks. These are part of the Chiltepines Member of the Potoikam Formation (see the map). They are are mostly siltstones and fine-grained sandstones that were originally deposited in a generally low-energy river that flowed through the area.

Look for the features depicted in the photos on this page.

Continue down the trail, following the stream and passing cliffs of dark weathered rhyolite dike rock until you return to the Jantzen House where this section of the log ends.

Three views of the Chiltepines Member.
Above: sedimentary layers deformed into an open fold.

Below: zones in the fine-grained rock that are populated with pebbles. A study of the pebbles indicates that they did not come from the local area and may have come from the northeast.
Right: layers of nodules or concretions that developed by concentration of chemical cement as the sediment was becoming a rock .

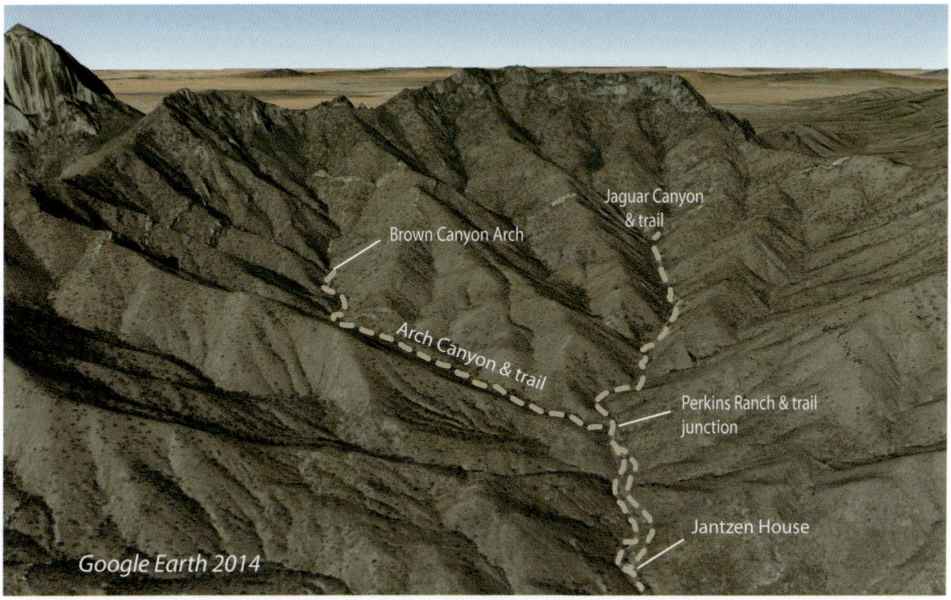

A Google Earth image showing the uppermost part of Brown Canyon. Rock-Walk 1 starts at the Environmental Education Center, passes the Jantzen House and ends at the trail junction at Perkins Ranch. Rock-Walk 2 continues up Arch Canyon and Rock-Walk 3 continues up Jaguar Canyon.

ROCK-WALK 2: Arch Trail, 1.4 miles (2.3 Km) to arch

Starting at the cattle chute and tank, walk up the main canyon. Near the windmill the trail enters a juniper-oak woodland. After some distance the trail turns sharply to the right, then left, as outcrops of dikes begin to appear. Then, as the trail follows the bouldery channel, watch ahead for the magical appearance of the Brown Canyon Arch. As you stand underneath it and look up, imagine you are inside a rhyolite dike. Look closely at the high slabs and you will find pictographs created by native people and early ranchers.

Although it is called an arch, technically it is a natural bridge. It was created as running water, with the help of fractures and weathering, eroded and undercut a rhyolite dike that crosses the stream course.

ROCK-WALK 3: Jaguar Canyon Trail, about one mile (1.6 Km) to end of trail.

Start at the cattle chute and head north up the trail past the tank. Depending on the condition of the trail, trail-finding may be somewhat challenging. You are leaving the beaten path for the wilds and mysteries of Jaguar Canyon. It was just over the ridge to the north that the jaguar Macho B was located, leading to a tragic encounter between a rare big cat and human society in 2009 (Jack L. Childs and Anna Mary Childs, *Ambushed on the Jaguar Trail*, 2008).

Outcrops of dark grey uniformly featureless Chiltepines Member occur along most of the canyon bottom. Streams carry rocks out of what is a nearly inaccessible part of the range. Large boulders of the Mulberry Wash Formation occur in the canyon bottom along with boulders of Baboquivari Peak Perthite Granite.

In about 0.50 miles a large vertical dike forms the sheer east side of the canyon. The opposite slope of the canyon is underlain by Mulberry Wash Formation. The trail disappears soon after (12R E 0445922, N 35156730). Very large boulders of Mulberry Wash Formation occur in the wash near this point. Several show parallel orientation of clasts resulting from either sedimentary layering or debris flow structure. Further exploration of the canyon is reserved for the well-prepared hiker.

Above: a dike of Allison Camp Rhyolite exposed in the upper part of Jaguar Canyon.

Left: Jaguar Canyon seen looking north from the trail junction. The area in foreground is underlain by the Chiltepines Member. Craggy outcrops are Allison Camp Rhyolite dikes. On the skyline is the Baboquivari Peak Perthite Granite. The darker rocks below the ridge are the Mulberry Wash Formation.

Right: boulder of Mulberry Wash Formation showing alignment of clasts. In this view they are aligned vertically above the walking stick.

References and Suggested Reading

Before this guide was written, there was nothing non-technical published on the geology of the Baboquivari Mountains and Brown Canyon. A compilation of part of the Tohono O'odham Nation geologic work contains a helpful map and is presented in Beikman, et.al, 1995, *Geologic map of the Tohono O'odham Indian Nation, Southern Arizona,* USGS Geologic Survey Miscellaneous Investigations Map I-2017, scale 1:125,000.

Gordon Haxel's relevant recent work has focused on Jurassic and Cretaceous geology of the area. The following include discussions of some Brown Canyon rocks:

Haxel, et al., 2005, *Middle Jurassic Topawa Group, Baboquivari Mountains, south-central Arizona: volcanic and sedimentary record of deep basins within the Jurassic magmatic arc,* in The Mojave-Sonora Megashear Hypothesis: Development, Assessment, and Alternatives: Geological Society of America Special Paper 393, p. 329–357.

Two helpful papers by Haxel are in Spencer, J.E., and Titley, S.R., editors, 2008, *Circum-Pacific tectonics, geologic evolution, and ore deposits*, Arizona Geological Society Digest 22.

Haxel, G.B., et al., 2008, *Geology and Geochemistry of Jurassic Plutonic rocks, Baboquivari Mountains, South-central Arizona*, in Arizona Geological Society Digest 22, 2008.

Haxel, G.B., et al. 2008, *Late Jurassic igneous rocks in south-central Arizona and north-central Sonora: magmatic accompaniment of crustal extension*, in Arizona Geological Society Digest 22, 2008.

A paper published on the deformation and metamorphism of rocks that are found in Brown Canyon and also occur a few miles to the south is: Goodwin, L.B., and Haxel, G.B., 1990, *Structural evolution of the southern Baboquivari Mountains, south-central Arizona and north-central Sonora*: Tectonics, v. 9, p. 1077–1095.

A brief description of the Brown Canyon area geology included in a study prepared for the Baboquivari Peak Wilderness Area study is Nowlan et al., 1989, *Mineral Resources of the Baboquivari Peak and Coyote Mountains Wilderness Study Area, Pima County, Arizona*, U.S. Geological Survey Bulletin 1702-E.

The Arizona Geological Survey (http://azgs.az.gov/) is an excellent source of information on the region. While they have nothing specifically on the Baboquivari Mountains, there are many technical and popular publications on various places in Arizona. A favorite is the Arizona Geologic Highway Map. Also check out their *Down to Earth* series. Of course, they have a Facebook Page and host several excellent blogs.

The best place to start any pursuit of the natural history of the region is the Arizona Sonoran Desert Museum and bookstore. Begin with an outstanding and comprehensive book published in 2000 titled *A Natural History of the Sonoran Desert*, edited by Steven J. Phillips and Patricia Comus. The museum website contains a recommended overview of Sonoran desert geology written by Robert Scarborough: http://www.desertmuseum.org/-books/nhsd_geologic_origin.php.

The Sonoran Desert Museum has also published *A Natural History of the Santa Catalina Mountains, Arizona* by Richard C. Brusca and Wendy Moore, 2013. Although its focus is the Santa Catalina Mountains, it is very relevant to the entire region. Chapter One: *Geologic Forces Shape the West* and Chapter Two: *Islands in the Sky, An Introduction to the Madrean Sky Islands* can be considered essential reading for understanding the Baboquivari Mointains.

The paleogeographic maps used in this book are produced by Ron Blakey of Colorado Plateau Geosystems. His website, http://cpgeosystems.com/, contains a wealth of maps of the ancient Earth. Be sure to check out the gorgeous book he wrote with Wayne Ranney titled *Ancient Landscapes of the Colorado Plateau.*

The following papers were used in the Landscapes Chapter:

Menges, C. M., 1981, *The Sonoita Creek basin: Implications for late Cenozoic Tectonic evolution of basins and ranges in south-eastern Arizona* (unpublished M.S. thesis): University of Arizona, Tucson.

Menges, C., and P. Pearthree (1989), *Late Cenozoic tectonism in Arizona and its impact on regional landscape evolution, Geologic evolution of Arizona*: Arizona Geological Society Digest, 17.

Jungers, M.C., 2012, *Post-Tectonic Landscape Evolution in Southeastern Arizona: When Did a River Start to Run Through It?*, Arizona Geology Magazine, Winter 2012.

Richard, Stephen M., Shipman, Todd C., Greene, Lizbeth C., and Harris, Raymond, 2007, *Estimated Depth to Bedrock in Arizona*: Arizona Geological Survey Digital Geologic Map DGM-52, version 1.0, map scale 1:1,000,000, 9 p. and 1 map sheet.

The literature on the geology of Arizona is as rich as the geology itself and there is an amazing world awaiting the determined student.

GLOSSARY

aggradation	The process of a stream building up its flood plain level by deposition.
albite	A *plagioclase feldspar* with the formula $NaAlSi_3O_8$. It is white, light grey and can exhibit precisely inscribed lines on a cleavage face. It also displays the oblique *cleavage* of all *plagioclase* feldspars. Both properties help distinguish it from *orthoclase*.
alkali feldspar	A *feldspar* that contains the alkali elements sodium and potassium.
alluvium	Unconsolidated clay, silt, sand, and gravel deposited on the Earth's surface by streams, wind, or gravity.
arroyo	A deep, steep-sided, flat-floored channel of an intermittent or ephemeral stream. A feature of arid and semiarid areas. The term is most commonly used in the southwestern United States
bajada	A broad slope of alluvium extending from a mountain front into the adjacent basin. It is formed by the coalescence of confluent *alluvial fans*. It is an arid and semi-arid feature typical of the southwestern United States.
ballenas	A rounded elongate ridge, shaped like a whale's back, that is formed by the dissection and extensive erosion of an *alluvial fan*. Ballena is the Spanish word for whale.
basement	The complex of *igneous* and *metamorphic* rocks that occurs beneath *sedimentary* rocks. Most often Proterozoic or older, but can refer to younger crystalline rocks beneath a more recent sedimentary and volcanic cover.
Basin and Range Province	A region of the inter-montane west characterized by north-south trending subparallel fault-bounded mountains separated by sediment-filled valleys. The topography, best seen in Nevada, extends across parts of Oregon, Idaho, California, Utah, Arizona, and New Mexico into Mexico. The features formed during extension of the crust from about 30 Ma to 5 Ma.
biotite	A black, dark green, or dark brown mica mineral composed of potassium, magnesium, iron, aluminum, and silica. Biotite is one of two common *mafic* minerals in *igneous* and *metamorphic* rocks. The other is *hornblende*.
breccia	A coarse-grained clastic rock composed of angular broken fragments held together by cement in a finer grained matrix. It can result from sedimentary processes with minimum transport and wear, igneous process like explosive ejection of material, or tectonic processes like fault *breccia*.
clast	A grain, fragment or other constituent of a sediment produced by the breakdown of a previous rock.
cleavage	The systematic breakage of a mineral along planes of weak atomic bonds. It can yield precise reflective planes in the broken mineral grain. The presence of cleavage is helpful in identifying certain minerals.

conglomerate	A coarse-grained clastic *sedimentary* rock composed of rounded to subangular *clasts* held together by cement in a finer grained matrix. It results from sedimentary processes and is the consolidated equivalent of gravel.
contact	The interface between adjacent rock bodies. It can be a depositional, intrusive or a fault contact.
cross-cutting relationships	A principle of geology that states that a geologic feature which cuts across or interrupts the pattern of another is the younger of the two features.
degradation	Generally the wearing down of the Earth's surface by weathering and erosion. Specifically the process of a stream lowering the level of its channel by cutting downward to maintain its *grade.*
depositional environment	A place with certain physical, chemical and biological processes where a sediment is deposited. Comparing ancient sediments with modern sediments and their environments is important to understanding and reconstructing the history of ancient sediments and their environments. The *paleogeographic* maps in this guide rely on such studies.
dike	A discordant (cuts across the structures and features of the host rock), tabular *intrusive* body.
eolian	Processes and deposits resulting from the action of wind.
epidote	A calcium, aluminum, iron, silicate mineral formed by alteration of other minerals under low and medium grade metamorphism. Epidote occurrences in the Brown Canyon area are better called products of alteration because the conditions of change are primarily chemical.
extension	A term that implies an increase in length of an object under tension or stretching. Also refers to the process of lengething through tension and stretching.
extrusive	*Igneous rocks* that have been formed from *magma* extruded onto the Earth's surface as well as the process of extrusion. Synonymous with *volcanic*. As opposed to *intrusive.*
fan (alluvial)	An open fan-shaped or cone-shaped deposit of *alluvium* that has been eroded from a mountain valley and distributed outward from the base of the mountain where the stream issues from the valley and its gradient suddenly decreases.
fault types	Normal fault: body of rock above fault (hanging wall) slides down under tensional stress. Reverse fault: hanging wall rock is pushed up under compressional stress. Strike-slip fault: bodies move horizontally parallel to *strike* of fault.
feldspar	A group of rock-forming minerals that make up 60% of the Earth's crust. They are aluminum *silicates* most commonly containing sodium, potassium, and calcium.

felsic	A term applied to rocks containing an abundance of light-colored minerals. Also refers to the magma that would produce such a mineralogy and to the minerals *quartz, orthoclase, and plagioclase*. It implies a silica-rich chemistry.
foliation	The preferred alignment of stuctures or features in a rock. It can be bands and streaks of different mineral composition, alignment of elongate or platy (flat) crystal grains, or splitting of rock into thin sheets. A common feature of *metamorphic* rocks.
formation	A geological formation is a readily recognized and mappable set of associated rocks such as sedimentary beds, intrusive rocks, or volcanic deposits.
glacial (stage)	The interval of time during the growth and disappearance of ice sheets. The Pleistocene Epoch is subdivided into glacial stages corresponding to times of colder climate and advance of glaciers.
graben	A subsided trough or basin bounded on both sides by *normal faults*. If the basin is bounded by only one fault it is called a half-graben. The basins of the *Basin and Range Province* can be both. The up-thrown counterpart is called a *horst*.
grade (stream)	A condition of a stream in equilibrium between erosion and deposition. It is ". . . one in which, over a period of years, slope is delicately adjusted to provide, with available discharge and with prevailing channel characteristics, just the velocity required for the transport of the load supplied from the drainage basin." (Mackin, 1948).
granite	A coarse-grained *plutonic igneous* rock composed primarily of *quartz* and *feldspar*, commonly with minor accessory *mafic* minerals like *hornblende* and *biotite*.
granodiorite	A coarse-grained *plutonic igneous* rock containing *quartz, plagioclase, alkali feldspar,* and *mafic* minerals. It is less *felsic* than *granite* and the *mafic* minerals are more apparent.
hornblende	A black, dark brown, or dark green *silicate* mineral containing calcium, sodium, magnesium, iron, and aluminum. It is a common *mafic* mineral in *igneous* and *metamorphic* rocks along with *biotite*.
horst	A relatively high fault block bounded by normal faults that is adjacent to a *graben*. The ranges of *the Basin and Range Province* could be considered to be horsts. They are high relative to the adjacent blocks because they have been uplifted or remained stationary as the adjacent block subsided.
hypabyssal	*Igneous* rocks that are emplaced in the crust at shallow depths. They may exhibit characteristics of lava flows because of rapid cooling and may be transitional between *plutonic* and *volcanic* rocks. *Dikes, sills,* and feeder pipes for volcanic activity are hypabyssal. Synonymous with *subvolcanic*.
igneous	Rocks that have formed by the cooling and crystallization of *magma*.

interglacial (stage)	The interval of time between *glacials*. An interglacial is a time of warmer climate and reduced occurrence of glaciers and ice sheets in lower latitudes. Similar to present conditions.
intrusive	A term that refers to a body of *igneous* rock that has been emplaced within the crust and also the intrusion process. Synonymous with *plutonic*. As opposed to *extrusive*.
Laurentia	The name given the ancestral North American continent. Its rocks are contained within the continent's core (craton) and are Archean in age. During early Proterozoic it was the core of the supercontinent Nuna which broke up and reformed as Rodinia. It appeared as a separate continent after rifting from the Rodinia supercontinent about 750 Ma.
leucocratic	A very light-colored (*felsic*) *igneous* rock. *Mafic* minerals are very minor or absent.
lithosphere	The rigid upper layers (crust and upper mantle) of the Earth that float on and glide across the partially molten layer below it called the asthenosphere.
Ma	The abbreviation for millions of years ago.
Madrean Sky Islands	Sky islands are high mountains separated from one another by intervening basins and they characterize the landscapes of southeast Arizona, southwest New Mexico, and northern Sonora. The contast in ecosystems is sufficient to isolate the mountains from one another, preventing the free movement of woodland and forest species. The mountains are high enough to host oak-woodlands and the basins are desert and desert grasslands. Madrean refers to the Sierra Madre Occidental mountains of western Mexico and the Madrean pine-oak woodlands that are used to define the sky islands ecologically.
mafic	A term applied to rocks containing an abundance of dark-colored minerals. The term also refers to the magma that would produce such a mineralogy. It implies a silica-poor chemistry. As opposed to *felsic*.
magma	Molten or partially melted rock, primarily *silicate* material. It is the source for *igneous rocks*.
magmatic arc	An arcuate (curved) line on the Earth's surface of igneous activity expressed as plutons and volcanic centers. It is formed from magma created by the melting of upper mantle and crustal rocks above a subduction zone.
metamorphic	Rocks that have changed from the original parent rock because they became unstable under new temperature, pressure, and chemical conditions.
numerical age	The age of an event expressed in years. Preferred over synonym *absolute age*. Contrast with *relative age*.
orthoclase	A kind of potassium *feldspar*. It is opaque white, cream yellow, flesh-pink, or grey and exhibits a near right-angle *cleavage*. The most common mineral in *granite*.

paleogeography	The reconstruction of ancient landscapes and geographies using the occurrences and historical changes of different rock types, structures, depositional environments, plate tectonics, fossils, and paleomagnetism.
Pangea	A supercontinent that existed during the late Paleozoic and early Mesozoic eras. It was assembled from earlier continental units about 300 to 250 million years ago, and it began to break apart about 175 million years ago.
perthite	A *feldspar* that is composed of very thin intergrown lamallae of *albite* and *orthoclase*. If the lamallae are large enough to be visible they can be seen as white strands of albite in the pink host, potassium *feldspar,* which is called microcline.
piedmont	Generally a plain or slope at the base of a mountain. In the arid and semi-arid *Basin and Range Province* it is a broad slope of *alluvium* extending from a mountain front into the adjacent basin. It is formed from the coalescence of confluent *alluvial fans*. Synonymous with *bajada.*
Pinal Schist	An assemblage of metamorphic rocks that occurs in southern Arizona and is 1.65 billion years old. It is interpreted as being part of a subduction zone complex that developed along the southern edge of the North American continent. These are the oldest rocks in southern Arizona.
piracy (stream)	The diversion of the drainage of one stream into the drainage of an adjacent stream. This can happen when headward erosion of a stream breaches the divide between them and captures the adjacent drainage. Capture can also happen when a basin overflows and spills into an adjacent basin. Drainage of the channel can then become the drainage of the captured basin.
plagioclase	A group of *feldspars* varying continuously in chemical composition from Na-feldspar (*albite*) to Ca-feldspar (anorthite). They commonly display near-microscopic twinning lamallae that appear as precisely inscribed lines on a *cleavage* face. Oblique cleavage is diagnostic.
plate tectonics	The theory of global tectonics in which the Earth's *lithosphere* is separated into rigid plates that move and interact with one another, creating the tectonic features seen in the *lithosphere.*
pluton	A body of igneous rock that has crystallized from magma which has intruded into the surrounding rock. Plutons include *batholiths, stocks, dikes, sills,* and other *intrusive igneous* bodies.
porphyritic/porphyry	A porphyritic texture exhibits significantly larger crystals imbedded in a ground mass of much smaller grains. The texture indicates an earlier period of cooling and crystallization when the crystals were able to grow large, followed by cooling and crystallization of the remaining magma under conditions that inhibited growth. Porphyry is the name for rocks that exhibit porphyritic texture.
pyroclastic flow	A high velocity density current of hot ash, gas, and rock resulting from a very explosive volcanic eruption.

quartz	The *silicate* mineral with the formula SiO_2 (silica). *Silica* is the fundamental chemical of the Earth's crust and quartz is a major component of *granite.*
relative age	The age of an event or rock relative to the age of other events or rocks. Contrast with *numerical age.*
rhyolite	A fine-grained *igneous* rock that is the chemical equivalent of *granite.* Occurs as *volcanic* and *subvolcanic (hypabyssal)* rocks.
sedimentary	One of the three fundamental types of rocks. It is composed of sediment deposited on the Earth's surface.
silica/silicate	Silica refers to minerals with the formula SiO_2. It is most abundant as the mineral *quartz.* It also occurs as cryptocrystalline chalcedony, agate, and non-crystalline substances like opal. Chert, flint, and diatomite are less pure forms of silica. Silicate is the name applied to minerals with silicon and oxygen in their chemistry.
sill	A tabular *pluton* that intruded parallel (concordant) to the structures and features of the host rock.
stock	A *pluton* with a surface exposure of less than 40 sq mi (100 km^2). Stocks are more massive than tabular plutons.
strike and dip	The method for measuring a planar feature's attitude in space. Strike is the compass bearing of a horizontal line in the plane and dip is the angle of tilt from the horizontal of the feature perpendicular to the strike. The information is shown on a map using a standard symbol accompanied by the dip value (⊥ 25).
subvolcanic	*Igneous* rocks that are emplaced in the crust at shallow depths. They may exhibit characteristics of lava flows because of rapid cooling and may be transitional between *plutonic* and *volcanic* rocks. *Dikes* and *sills* and feeder pipes for volcanic activity are subvolcanic. Synonymous with *hypabyssal.*
subduction zone	A plate boundary where one *lithospheric* plate descends below the edge of another. The descending plate can carry ocean floor material and water to sufficient depth to cause melting of upper mantle and lower crust. *Magmatic arcs* occur above the zone of melting. Significant seismic activity accompanies subduction.
tectonic	Refers to the dynamics of the Earth that influence the evolution of the crust, the processes and products of deformation, and the production of crustal structures like faults, folds, volcanic features, mountains, continents, and ocean basins. The concepts of plate tectonics help relate and explain geological processes and features.
terrace (stream)	A step-like surface elevated above a floodplain or another terrace. The terrace surface was an earlier floodplain that was incised by the stream as it cut downward in response to changing conditions.

tuff	A rock formed by the lithification of volcanic ash and other relatively fine-grained *pyroclastic* ejecta. If it was still hot and viscous when it landed, it could form welded tuff.
volcanic	*Igneous* material that is erupted onto the Earth's surface. It can vary from coherent lava flows to explosively ejected fragmental material including ash.
volcaniclastic	A deposit of *clastic* volcanic material that has fragmented and been dispersed by water, wind, or gravity.

FIELD NOTES

FIELD NOTES